Improving the flow situation in the building

Ava Hashempour

Title: Improving the flow situation in the building
Author: Ava Hashempour
Publisher: American Academic Research, USA
Cover designer: Ali Khiabanian
ISBN: 9781947464223

Introduction:

This study focuses on the concept of sustainable development in the engineering literature. Many disciplines such as building engineering, air conditioning, architecture and planning deal with this issue. The dismantling of the frontiers of pure science in the world and the production of interdisciplinary sciences in the world is developing. This study is also conducted as an interdisciplinary form in architecture and air conditioning and with the help of approaches and techniques of museum management in the world, would able how to use different sciences to help stabilize the temperature of museums (and other buildings). In the first chapter, the subject literature deals with the interiors atmosphere of buildings by using existing techniques related to the background in the world. In the second chapter, we will introduce and design of the museums. Finally, in the third chapter, we will display experiences in one of the historical museums in order to show an achievement in building engineering literature.

Respectful - Ava Hashempour2020

Table of Contents

Table of Figures

List of tables

List of charts

Chapter One:

Energy optimization in the building

1.1 Introduction

Due to the reduction of fossil fuels, air pollution and global warming, the issue of air conditioning and improving indoor air flow is very important. Regarding high cost of air conditioning equipment, as well as the high energy consumption of these devices, their use does not help much in reducing energy consumption. In this part, such issues can be the responsibility of architects and building engineers with regarding to their design and interference in building elements such as openings, walls, etc., create this isothermal. In this project, isothermal is important from two aspects:

1. In order not to waste energy
2. In order not to damage the objects inside the building, which is very important to maintain

1.2 Statement of Problem

In recent decades, due to the rapid development of knowledge and engineering sciences, the need for multiple specialties in design is more than before, and more aristocracy of architects, engineers and urban planners to engineering issues in design is not enough. Today, designers must be able to improve the performance of their building with new scientific methods. One of the issues that is raised today is the temperature balance inside the building that each building has a different temperature depending on its use. In these studies, the balancing of interior spaces is discussed. According to existing standards, indoor spaces should have a temperature of 22-23 centigrade and this flow should be perfectly uniform throughout the building.

According to the observations, the temperature difference between different parts of the interior spaces is large and even this difference between 6-8 degrees has been observed. In addition, based on observations, indoor air flow where it is close to the ventilation system is good and it is less in other places. These problems are caused by improper placement of building components. The aim of this research is to place architectural variables in interior spaces in a way that it provides us with the further reduction of temperature difference and uniformity of indoor air flow; because if the indoor spaces are not at the isothermal, it will cause damage to objects, human health and heat dissipation. According to the standards, the indoor temperature should be around 23 degrees, which is done by ventilation systems but the main purpose of this study is the standardization of indoor air temperature, which is strongly dependent on architectural parameters and variables.

In these studies, the independent variables are the architectural elements such as: dimensions of the height and position of the openings and windows and the air temperature is dependent variable. The main challenge in this research is that because the air flow is pleasant and uniform and the difference in temperatures in different seasons and times of the day and the humidity in the air ... have a direct impact on indoor space and this causes destruction in the long and med term.

1.2.1 What is Air Conditioning?

Air conditioning is the act of replacing and moving air in a space by mechanical or natural means. Ventilation is often done by moving the indoor with the outside air.

In general, natural ventilation in the building has three different functions, which are:

- Providing breathable air inside the building by replacing fresh outside air with polluted and consumed indoor air. This function is also called ventilation for health.

- Creating physical comfort by increasing the rate of reduction of excess body temperature by evaporation of sweat created on the skin. Also by relieving the discomfort caused by wetting the surface of the body with sweat. This function is also called ventilation for comfort.
- Creating physical comfort inside the building by cooling the building material when the indoor air is warmer than the outside air. This function is considered as ventilation to cool the building (Rahaei & Azemati, 2019).

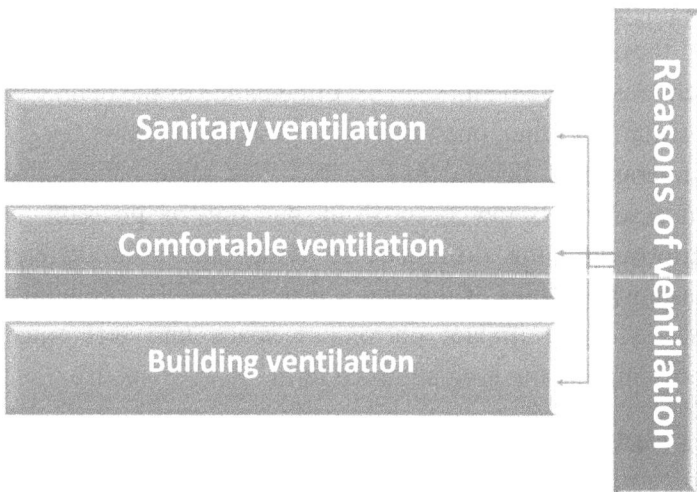

Sanitary ventilation

Comfortable ventilation

Building ventilation

Reasons of ventilation

Digram 1. Natural ventilation performance in the building

Conventional air conditioning methods are usually done in the following two general ways:

Air Conditioning: Artificial and natural ventilation. But more precisely, ventilation systems can be divided into the following categories:

| Natural ventilation |
| Natural ventilation with mechanical devices |
| Mechanical ventilation systems (ducted) |
| Air conditioning systems |
| Combined systems |

Classification of ventilation systems

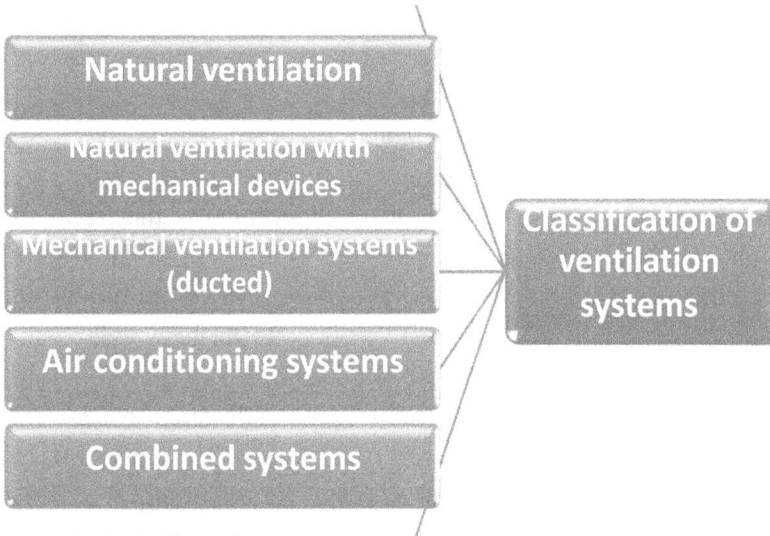

Digram 2. Classification of ventilation systems

1.2.2 Natural ventilation and its principles

Natural ventilation can provide fresh air, help transfer internal odors and heat, cool the structure and reduce structural radiation, as well as create evaporative cooling of the body and air. Natural ventilation occurs for the following reasons:

- Wind air pressure difference
- Temperature air pressure difference

The benefits of natural ventilation are convincing: energy costs are dramatically reduced, air quality improves, and airborne chemicals are minimized by air conditioners or other mechanical devices. Overall, the use of natural ventilation in the home can have a very positive impact on the occupants, the building itself and the environment.

1.2.3 Natural ventilation mechanism

Wind is an effective and determining factor in the movement of air inside the building. When the wind hits the building, the direct flow of air around and above it is broken and scattered. In this case, the air pressure is high on the windward surfaces (pressure zone) and very low on the backward winds (suction zone). Therefore, pressure differences occur at different levels of the building. When the wind forms a rectangular structure vertically, the front walls are subjected to pressure and the back walls are subjected to negative suction or pressure. If the wind blows obliquely into the building, the two opposite surfaces will be pressurized and the other two will be sucked.

The roofs of buildings are always located in the suction area. Of course, in the case of sloping roofs, this is true when the wind slope is low. Windward surfaces of steep slopes are located in the pressure zone and their backward wind surfaces are in the suction zone.

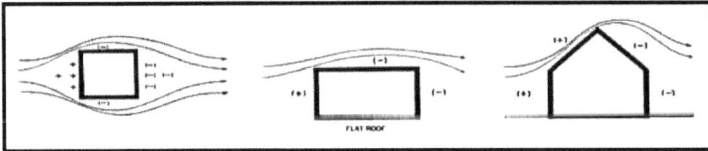

(Figure 1). Sloping roof pressure and suction, flat roof wind pressure and suction and the effect of wind on the building

The pressure difference that is created in this way on the surface of the building walls can be used to create natural ventilation and draught inside it.

1.2.4 Use of natural ventilation in the past

The importance of wind in the design and construction of residential environments has long been considered. Four centuries BC, Aristotle and Russian architect Vitruirus one century BC discussed about using wind in architecture and urban planning. In our country, for many centuries, all buildings have been built according to the climate and environmental conditions. Sun, wind, humidity, cold, heat and in general climatic and geographical conditions have had a direct impact on the traditional architecture of Iran in different regions. For example, to create a cool atmosphere inside warm desert houses, windcatchers, which is an innovative Iranian method, were used to make the living space bearable, in the following, we will discuss how it works:

1.2.5 New methods of using natural ventilation

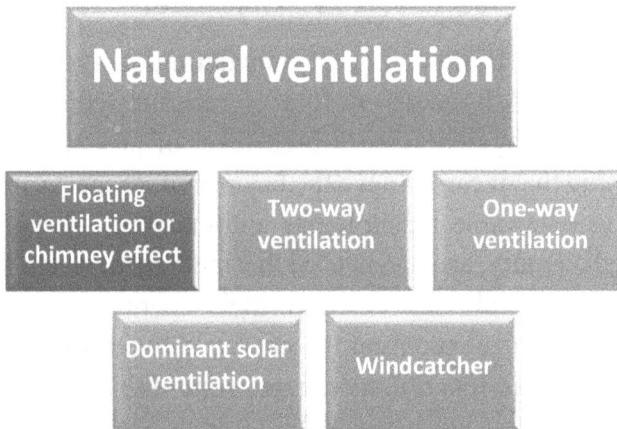

Natural ventilation

- Floating ventilation or chimney effect
- Two-way ventilation
- One-way ventilation
- Dominant solar ventilation
- Windcatcher

(Figure 2). Classification of ways to use natural ventilation

1.2.6 Windcatcher

Windcatcher is an innovative Iranian method for creating a cool atmosphere inside hot desert houses. This air conditioner has made the living space of the Iranian people bearable for many years. Windcatchers are usually small turrets in the form of quadrilaterals or regular polygons, which are mostly higher than other parts of the house and are located on the roof. Windcatchers are generally built on a part of the desert houses called the house-pond. The house-pond was a small porch located at the end of the summer rooms of each mansion. The windcatchers were located just above the pond and directed the air flow through the pores to the pond water. Existing studies in the field of natural ventilation of buildings show that the space of inlet and outlet openings of the windshield constitutes 3% -5% of the floor space and when there is no air, the windcatcher acts as a suction device.

1.3 Subject literature

1.3.1 Principles of Zero Energy Architecture

A Zero Energy Building (ZEB) is by definition a residential or commercial building with a significant reduction in energy needs by achieving a high level of productivity in such a way that it can achieve the balance of its energy needs by using renewable technologies and energies. In the late twentieth century, architects used this method to design low-consumption buildings to meet their energy needs and without being dependent on the energy network to send 25 k/m2 of surplus energy to the network during the year. Explaining these goals and defining them is critical to the design process (Torcellini, Pless, and Deru, 2006). During the last twenty years, about 200 residential and office-commercial projects in the world have been designed and built based on this new achievement and this trend continues to increase with rising energy prices. At first they had only a research aspect but they were became very soon as the path of public construction and the building market. The initial tendency to build this house was to reduce energy consumption and pay full attention to the local ecosystem (figure 3).

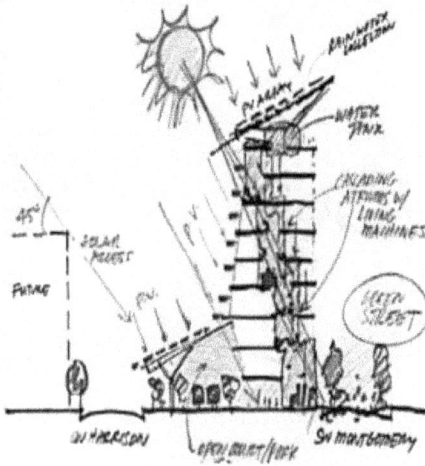

(Figure 3). Schematic diagram of the environmental performance of an office commercial building. Resourceses (http:// architecture-view.com/2010/06/06/Oregon-office-tower-zero-energy)

In this approach, the form of defining the goals and performance of a complex zero energy system and understanding the combined methods of applied productivity and renewable energy supply and consumption options is crucial. Inside the term zero energy is the idea that buildings can meet all their energy needs through renewable, clean, in-house, low-cost amounts. For the most stringent purposes, a zero energy system, such as a building, produces enough renewable energy in proportion to the annual energy consumption of the inhabitants. While in this type of buildings, strategies for conversion and change of energy sources, distribution and consumption are appropriate to the changing conditions of the building around the clock and climatic conditions (Figure 4).

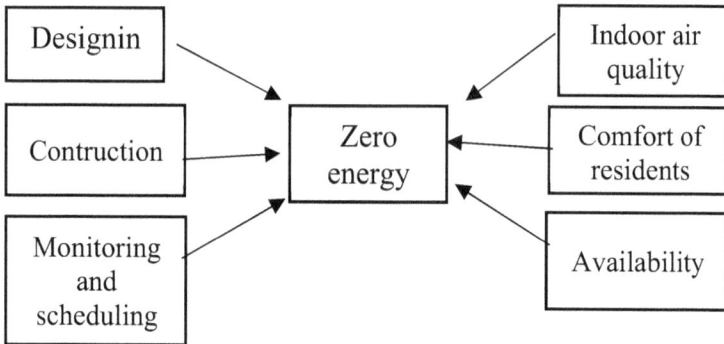

Figure 4. Communication diagram of zero energy house functions

Zero energy buildings self-sufficiently use both active and inactive capabilities simultaneously. In addition, the intelligent system monitors all stages of absorption and distribution, consumption and thermal regeneration as well. Technologies in active sectors can include: Application of photovoltaic cells, solar hot water, wind turbines, hydroelectric and biomass and supply of indoor air through subterranean cavities and cooling through well water flow inside the roof and in the inactive section refers to heat transfer and passing ventilation, direct and indirect sunlight absorption, and shading in summer (Figure 5).

(Figure 5). GE's proposed zero energy sample for 2015

In ranking the use of renewable energy technologies in these buildings, ceiling photovoltaic panels and solar water heaters are the most widely used items. Vertical-axis wind turbines on the building and horizon-axis separately from the wind building form are also in the next rankings. Table 1 shows the application of these related methods and technologies (Net-Zero Energy Building, 2012).

(Table 1) Ranking of recurring renewable energy supply methods in zero energy buildings

Application at zero energy	Searchable items
Reduce on-site energy consumption through low-energy building technology	Daylight, high efficiency HVAC equipment, natural ventilation, evaporative cooling and so on
	On-site supply options
Use of renewable resources embedded in the building	PV and solar water heater and wind turbine connected to the building
Use the resources available near the building	PV, solar hydrothermal power supply, wind and hydro on the site away from the building
	Off-site supply options
Use renewable sources which are provided off-site	Biomass, wood boards, biodiesel such as ethanol, biogas produced from site waste and used to generate heat and electricity for the building

Building zero energy is the product of responsible participation and coordination to achieve a project with sustainable goals so that all groups involved in planning, design and implementation are aware of their role in achieving the set goal and be responsible for the set actions. Integrated design process introduced in 2004 in the

Department of Architecture and Design at the University of Aalborg in Denmark is a combination of architectural knowledge and all areas of building engineering in which building problem-solving capabilities are possible based on interfaces between areas such as design, performance, energy consumption, interior, technology, and construction with the participation of all relevant people. This interdisciplinary trend brings the different languages of building groups closer together and paves the way for design integration. In a comprehensive process, project objectives can be represented and examined in different parts of the design and enter the comments of each group in the initial sections of Ideation and Phase Zero and One (Hansen, 2005). In fact, in the sketch section, the ideas of the architect and other engineers are combined and evaluated. This stage is the critical time period in which the interaction of the ideal ideas of architects and fuzzy logics of engineers is observed and inferred. Also in this section, the employer's ideas are re-read and the requested corrections are provided. Many simulation software evaluates the ideas of the architect, among other specialties, with the approach of analyzing environmental conditions at this stage, using a suitable graphical environment. Software such as Design Builder and EcoTect, and the like, can greatly represent engineers' predictions in terms of cost, energy, consumables, internal comfort function, and external impact of the building. Some software can also determine the environmental conditions of the building and its pollution level. In the data collection section, all the information needed to determine engineering solutions and appropriate decisions is created. Data from site and climate conditions, neighborhood, land form, light and shading, prevailing wind, proposed materials and available technologies in the form of determined cost and structural conditions and installation needs are carefully considered. In the inference stage, it is always possible to review the initial idea and the previous sections and make corrections until the optimal conditions are obtained with the participation of all members of the design team. This process is documented from both qualitative and quantitative dimensions and provides basic data and important and determining strategies to groups of engineers (Figure 6). Finally, it can be ensured that the building can meet the criteria and standards of policy organizations such as fuel optimization (Singh & Rakesh, 2014). Therefore, comprehensive attention to the possibility of utilizing energy supply capabilities in the environment is a characteristic of zero energy buildings.

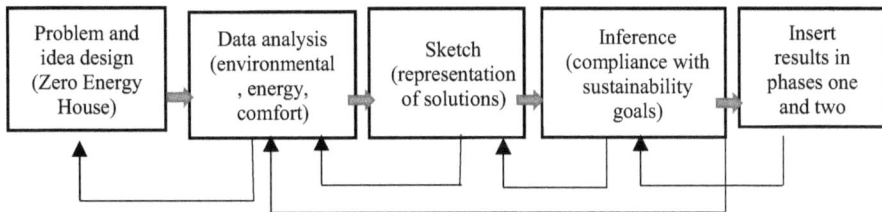

(Figure 6). Design diagram with a holistic approach in zero energy building

1.3.2 Investigation of energy sources

The sun is the source of all energy in the earth. In general, energy resources can be divided into two categories: renewable and non-renewable:

1.3.2.1 Non-renewable energy sources

Non-renewable energies can only be used once and their resources are limited and run out after a while, which are:
a) Fossil fuels
Includes gas and coal, such as oil and its products which formed over about five hundred million years and it has been

widely used as a source of energy by humans since the late 19th century. The most important problems of these fuels are polluting the environment due to the production of harmful gases such as carbon dioxide and ... their limited.
b) Nuclear fuels
c) Wood and firewood

1.3.2.2 Renewable Energy Sources

It is said that the energies that exist as far as the earth and the sun will be that have no destructive effects on the environment and they are in line with sustainable development. Some of these sources are as follows:
Solar energy;
Wind energy;
Water energy;
Wave energy;
Tidal energy;
Ocean thermal energy conversion;
Solar salt water pools;
Hydraulic energy: Hydropower using hydropower potential (hydropower plant);
Geothermal energy: consists of two words geo meaning earth and themral meaning heat. It refers to the heat and energy beneath the earth's surface;
Biomass: waste and discarded materials;
Hydrogen.
Benefits of using renewable energy:
1. no production of environmental pollutants, 2. free and unlimited energy source, 3. very long useful life, 4. easy access, 5. renewable resources.
Reasons for using renewable energy
The average temperature of the earth is about 33 degrees which had increased in the last one hundred years at the rate

of ./3 to ./6 degree. Pollution from human fossil fuels increases hazardous gases such as carbon dioxide, methane, etc. These gases cause a greenhouse effect. An increase in global temperature of 2 ° C compared to the current situation causes the average global temperature to rise above what has been experienced in the last 1000 years which have humans encountered such conditions at no time in their history. If the current situation continues, in every decade ./3 degree will increase the earth's temperature. Global warming could cause hurricanes in the atmosphere as well as changes in the warm flows of the oceans. 70 to 90% of carbon dioxide is obtained from energy consumption, especially from the combustion of fossil fuels. The remaining 10 to 30 percent is produced due to deforestation and land changes. The destruction of the ozone layer, which causes extreme heat and cancer, etc., as well as acid rain, are other disadvantages of excessive consumption of fossil fuels (www.mapnamd1.com).

1.3.3 Use of Solar Energy in Zero Energy Buildings

The strategies used in the construction of green buildings and zero energy are building design with a view to using solar energy which is often carried out with the aim of producing and optimal consuming energy. In these buildings, the position of windows, walls, porches, canopies and trees should be oriented in such a way as to create shade in summer and maximum solar gain in winter. In addition, the proper location of the window can increase the amount of daylight and reduce the electrical energy consumption of lighting during the day. The use of active and passive solar technologies, solar electricity, the use of green space on the roof of the building are among the effective solutions in this sector. In this project, architects have predicted and designed the building form and orientation of the building for shading in summer and absorbing heat in winter directly and indirectly (greenhouse). Facility engineers consider all possible solutions to meet the comfort needs of the building in the areas of lighting, cooling and heating, ventilation, wastewater recycling, consumables and their recycling based on life cycle.

1.3.3.1 The Nature of Solar Energy

The sun's surface temperature, equal to 6,000 degrees, has oxygen available to burners for billions of years and it is the only reliable source of energy for the planet. The sun sends more than 1,000 times more energy to the earth every day than humans need. According to calculations, the amount of energy that the sun sends to the earth in one hour is equal to the energy of 23 billion tons of coal. According to scientific estimates, about 4.5 billion years have passed since the birth of this ball of fire and it can still be considered a huge source of energy for the next 5 billion years. Annually, 1034×7.2 j radial energy reaches the earth's surface. The potential of solar energy in the world is so great that it exceeds the flow and future needs of total global energy demand (http://www.satba.gov.ir).

1.3.3.2 Justification of using Solar Energy and its capability in Iran

Approximately 1/1 in 10 to 20 kWh of energy is emitted from the sun every second. Only one billionth of this energy hits the outer surface of the Earth's atmosphere. This energy is equivalent to 1.5 in 10 to the power of 18 kWh per year. Due to reflection, dispersion and absorption by gases and suspended particles in the atmosphere, only 47% of this energy reaches the Earth's surface. Thus, the energy radiated to the earth's surface annually is approximately equal to 7 by 10 to the power of 17 kWh (http://www.satba.gov.ir).

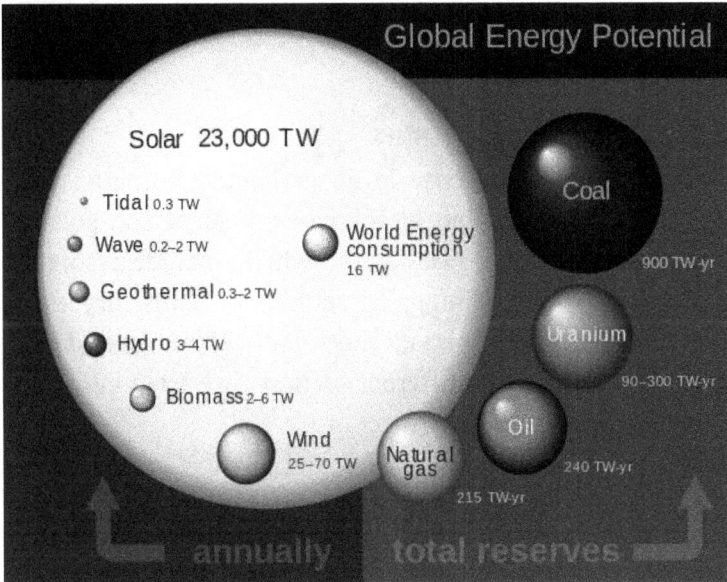

Global Energy Potential

Solar 23,000 TW

Tidal 0.3 TW

Wave 0.2–2 TW

World Energy consumption 16 TW

Coal 900 TW-yr

Geothermal 0.3–2 TW

Hydro 3–4 TW

Uranium 90–300 TW-yr

Biomass 2–6 TW

Wind 25–70 TW

Natural gas 215 TW-yr

Oil 240 TW-yr

annually   total reserves

(Figure 7). The amount of solar energy absorbed by the earth

The amount of solar radiation in Iran is estimated between 1700 to 2200 kWh per square meter per year, which is higher than the global average. In Iran, an average of 280 sunny days are reported and except for the Caspian coast, the percentage of sunny days per year is between 63 and 98%. In Iran, an average of 5.5 kilowatt hours of solar energy per square meter of the earth's suface. The area of Iran is approximately 1,600,000 square kilometers, which is about a square meter. The total amount of solar radiation during the day for Iran is approximately equal to 9 billion megawatt hours. If we absorb solar energy from only 1% of Iran's area and the efficiency of the energy receiving system is only 10%, then we can receive 9,000,000 MWh of energy per day from the sun. The total amount of energy that the earth receives from the sun is very high and is equivalent to the energy produced by burning 3 million tons of gasoline per second which shows that the use of solar energy in Iran is technically and economically practical (http://www.satba.gov.ir).

1.4 Solar Energy Applications

Applications of solar energy in general into two parts: 1. Power plant, which includes: solar power plant and solar chimney and 2. Non-power plant, which includes: solar water heaters, solar buildings, desalination plants, dryers, stoves, ovens, swimming pools and solar heat pumps and in the present study and in different sections, we examine the systems appropriate to the project process.

Thermal chimney

Thermal chimney is used to let steam and air flow out of the building. By placing an outlet in hot areas, air is drawn in to ventilate the building. Sunshades are designed to ventilate the exhausting heat of summer in southern rooms through upper vents. The lower vents of the house open with the north windows and the air inside the house comes out of the upper windows of the sunroom.

Solar chimneys are placed in places where the sun can heat them so that they can also heat the indoor air. One of the advantages of a solar chimney is its automatic control power. In this way, the warmer the day temperature, the higher the air temperature will automatically increase.

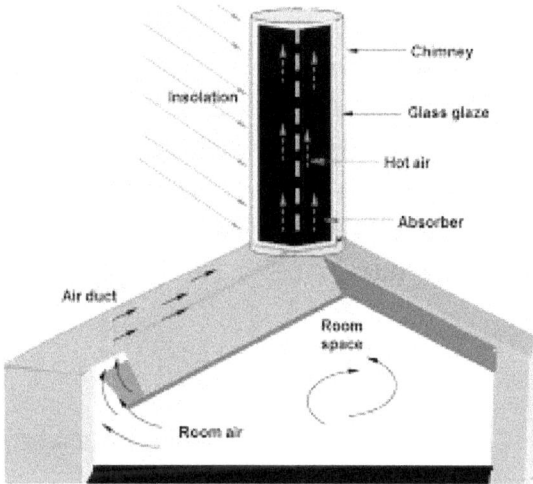

(Figure 8). Solar chimneys

Solar Water Heater

Water heaters are the main system used in non-solar solar applications used to heat water. This device is used to produce domestic hot water with a temperature of about 70 C. The solar energy can also be used to generate high temperatures, up to 3000C and above (sezhin.com).

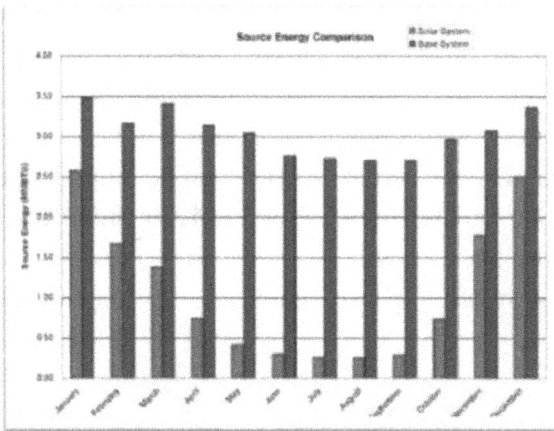

(Figure 9). Comparison of annual energy consumption for conventional water and solar water heater

Solar Desalinization

Unfortunately, Qom province has had salty and non-drinkable water since ancient times, and this issue has been one of the biggest concerns and problems of this province which has not yet been fully resolved. Using solar energy is a useful and effective way to desalinate drinking water consumed in this province.

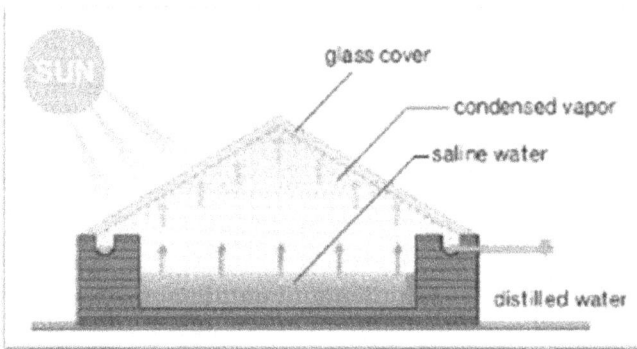

(Figure 10). Solar Desaliniztion

1.4.1 Solar heating & cooling (Solar buildings)

Supplying the thermal needs of buildings using the sun can be achieved in two ways: passive (passive-natural) and active (active). The quality and architecture of a building depends entirely on receiving and storing solar energy in the passive state, while in active form of solar heating, it requires the use of solar panels (Sheikh Pour & Mehmandoost, 2015). In the following, we will examine these systems, ie 1. Passive solar systems (passive) and 2. Active solar systems (active).

Passive solar heating system (passive)

In passive solar systems, buildings are designed to meet the needs of heating, cooling and lighting in a natural and climate-friendly manner and they are called "passive solar systems" because the need for heating and cooling equipment is minimized. A house designed with natural solar heating can reduce energy consumption by up to 75% while only 5 to 10% increase construction costs (Sheikh Pour & Mehmandoost, 2015). In passive solar systems, hot air of the southern part of the building by convective heat transfer, conveys throughout the building. Opaque materials absorb between 40 and 95% of the sun's rays; so, this amount was strongly dependent on the color of the objects. Therefore, the more turbid the body, the higher the absorption rate. For this reason, we try to make the sun-absorbing materials dark in color.

Types of passive solar heating systems (passive)

Passive solar heating is divided into two categories: 1. Direct reception and 2. Indirect reception, which themselves include the following:
1: Direct reception: a) direct sunlight enters the room through the windows (direct gain method), b) ceiling skylights;
2: Indirect reception: a) use of the solar energy storage wall of the Drum wall, b) water wall, c) heavy wall, d) use of a vertical flat receiver with natural air flow (solar chimney design), e) use the attached greenhouse, f) pool or pond on the water, g) Atrium, h) Thermosyphon phenomenon. In the following, we will briefly review these methods.
Direct receiving system

a) Direct receiving system (Direct reception): It is the simplest passive solar system. In this system, sunlight enters the interior space through windows, shutters and skylights and is absorbed by interior surfaces and furniture (Sheikh

Pour & Mehmandoost, 2015). Direct absorption buildings depend on south-facing windows (solar windows). Shortwave waves of solar energy pass through glass and enter space. Then, these waves heat the interior surfaces and emit long waves that are unable to leave the glass. In fact, these waves are trapped inside space and create a greenhouse effect (Sheikh Pour & Mehmandoost, 2015). The system consumes 40-75% of the solar energy directly hit by the window. In the system, the direct gain of walls and floors is considered a thermal mass. The amount of heat storage of different materials depends on their thermal conductivity, specific heat and density. Thermal conductivity often also increases with increasing density. As a result, materials with high heat capacity (concrete blocks, bricks, concrete, etc.) increase the performance of this system.

Advantages and disadvantages of direct gain method

The advantages of this method include: 1. Using natural light, 2. South view, 3. Low implementation costs, and 4. Passive heating supply.

Disadvantages of this method include: 1. Blurring with bright shadows with high contrast due to high light, 2. Disruption of privacy, 3. Problems due to manual operation of night insulation (Sheikh Pour & Mehmandoost, 2015).

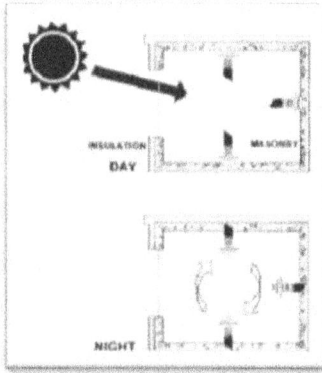

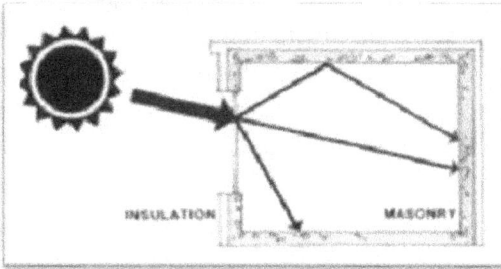

(Figure 11). Sunlight enters the building

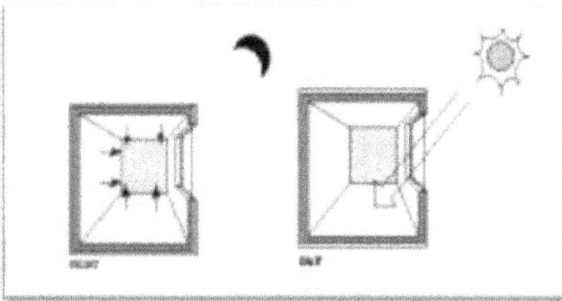

(Figure 12). Distribution of heat absorbed by sunlight in the building

Indirect gain (Absorption)

In indirect absorption, the reservoirs are a barrier between sunlight and the interior. It absorbs the sunlight that reaches and transmits to the home space. The system consumes 30 to 45 percent of indirect absorption of the solar energy that reaches the glass as a thermal mass. In the following, we will study these systems.

1.4.2 Drum Wall

This wall is located a short space from the glass and is made of high density materials such as stone, brick, clay or gallons full of water and their outer wall is dark. The space between the glass and the wall should be about 8 to 10 cm so that air circulation is easy. If single or double glasses are used, the window must be covered with thermal insulation at night so that the heat stored in the Drum wall is not easily transferred to the outside. Otherwise, triple glazing should be used. The heat stored during the day is gradually transferred into the building. Drum wall allows the sun to be used efficiently and its efficiency depends on the material, thickness and color of the wall surface.

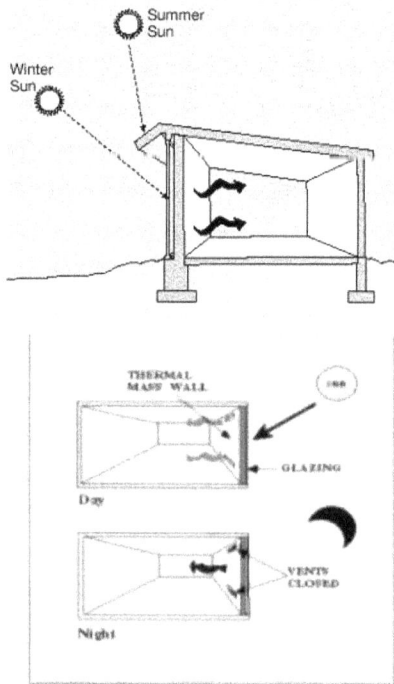

(Figure 13) Drum wall

Drum walls are generally used in places that do not require direct light, such as amphitheaters and the like or it is not possible to use the direct absorption system for some reason. In this way, hot air can be directed through the canal to different parts of the building, where it is not possible to use direct sunlight. It should be remembered that in summer, these walls are shaded and openings are installed on the windows until by opening them, hot air comes out between the glass and the wall. Due to the fact that in this system, the wall and the glass in front of it are very close, cleaning this glass from the inside is problematic considered one of the disadvantages of this system (memaran.ir).

1.4.3 Water Wall

Water wall is another type of Drum wall that instead of a solar wall with stone and clay materials, is made of metal barrels full of dark water that are placed in front of a sunny window. The heat storage capacity of water is twice that of heat mass. Therefore, the volume ratio is less required than the thermal mass. At least 13.23 liters of water per square foot of glass is poured into the tank. In this way, these walls can be made simply by stacking barrels of water next to each other horizontally or vertically, or smaller cans of water that are placed on checkered shelves. Water integrated wall is another type of such walls.

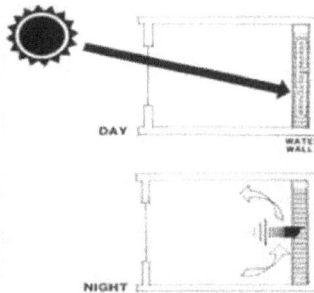

(Figure 14). Water wall

1.4.4 Rooftop pool or pond

In this method, the roof pool absorbs the heat of sunlight during the winter day and at night, the roof is covered with thermal insulation to maintain the resulting heat. In this way, the heat of the water inside the pool gradually keeps the air inside the building warm through conduction and radiation. The minimum depth of the water pool should not be less than 10 cm. Another type of this system is covered by glass. This system works differently in winter and summer. In hot climates, roof protection is very useful to prevent direct sunlight at noon (www.bernoulli.ir).

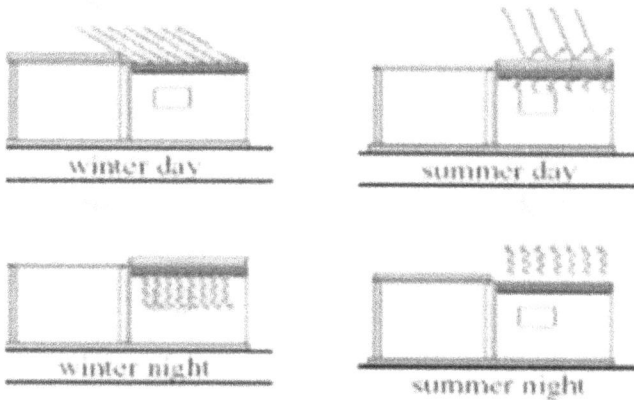

winter day   summer day

winter night   summer night

(Figure 15). Rooftop thermal pool or pond

1.4.5 Atrium

The atrium is a central space like a central courtyard in a building with a clear and sunny roof and different parts and spaces of the building are formed around the atrium space. Solar heat rays enter the atrium through a glass roof and in this way (greenhouse effect) thermal energy is stored in the atrium space. It enters the interior through openings and walls around the atrium. But what is important to prevent the spaces from heating up during the summer, the atrium space should be well ventilated and its glass roof should be effectively covered with appropriate canopies. Among the advantages of atriums are the function as a passive solar system in heating side spaces, the possibility of receiving adequate daylight, the possibility of optimal natural ventilation, proper access to side uses and creating a favorable architectural space in terms of vision and landscape. The most important drawback of atriums is the rapid spread of fire to the higher floors of the building.

The Phenomenon of Thermosyphon

Convective circulation of a fluid that occurs in a closed system where cold fluid replaces hot fluid in the same system is called a thermosyphon. This system is actually a natural displacement cycle. It should be noted that in the Drum wall (with valve) and the greenhouse, heat is distributed by thermosyphon as well. In this system, the energy absorption stage can be done connected to the building or in a completely separate environment and the heat absorbed by the channel is directed to the desired space and stored in a suitable place such as a concrete slab or stone warehouse, which is usually above the absorber surface.

Greenhouse method (greenhouse), isolated insulation (solar space), house insulation

The interface space in an input is the space between two in a row. In this way, in winter, this space prevents hot air from being directed outside when opening and closing a door. This space is also a good environment for lightening a lot of clothes in winter, clearing mud outside and getting rid of extra clothes before entering. A greenhouse can act as an interface space. In warm climates, this space is a good help to keep indoor rooms cool (Cook, 2018). Greenhouses work with a separate absorption system and are also called solar space (It is known as heat absorption and energy storage both in a separate space such as a greenhouse). A greenhouse or solar space adjacent to the main space of the house, with windows with shutters at the bottom and top can create natural ventilation in the building during the summer. When the windows open on sunny days, the heated air is allowed to escape through the upper open vents. Instead, colder air enters the space via the lower vents. This phenomenon is called the chimney effect and it can be used for natural cooling of solar buildings. The insulation system uses 15-30% of the sunlight that reaches the glass to heat the home space as well as maintains solar energy in solar rooms.

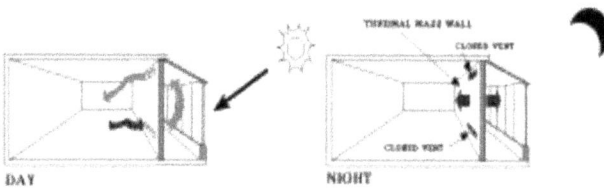

(Figure 16). How to receive, store and use solar energy in solar greenhouses - day and night. http://architectsofhoma.ir

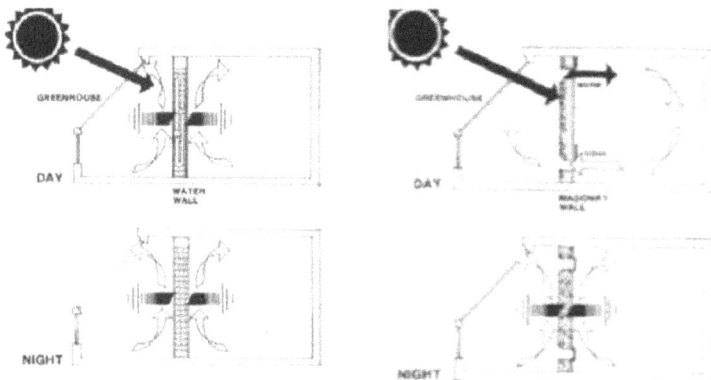

(Figure 17). Right image: Using an adjacent greenhouse with a Drum wall, Left image: Use the adjacent greenhouse with a water wall

1.4.6 Methods of Direct Passive Solar Energy

Passive solar design policies differ due to the location of the building and the climate of the area; however, the basic methods are the same. One of the most important of them is increasing solar heat gain in winter and decreasing it in summer.

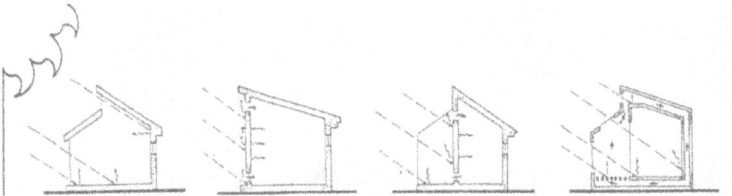

(Figure 18). A few examples of passive building heating systems (passives)

Special techniques include:
Start working by applying the strategy of optimal energy consumption plan.
Orientation of the longitudinal axis of the house to the east/west.
Build south-facing shades to provide window shade in summer and allow the sun to shine in winter.
Adding thermal mass to walls or floors to increase heat storage and not covering it with furniture.
Use of natural ventilation to reduce or eliminate cooling needs.
Use of daylight to provide natural lighting.
Orientation and estimation of the selected window type to optimize heat gain in winter and minimize heat gain in summer for a specific climate (Watson, 1998).

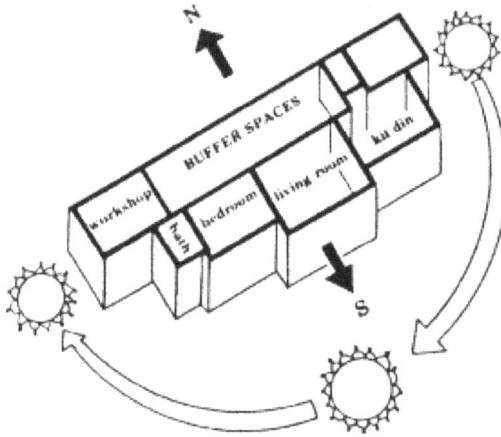

(Figure 19). Deployment of primary and secondary spaces in a solar house (The main spaces are on the south front and the secondary spaces are on the north front).

The longest wall of the house should be at an angle of plus or minus 15 degrees to the true south to get the most winter heat and reduce summer cooling costs. Deviation above that value results in lower winter temperatures than the optimum. At 30 degrees east or west south (true south) the winter heat gain is up to 15% lower than the optimum. Shrinking east and west walls and windows reduces excessive heat in summer.

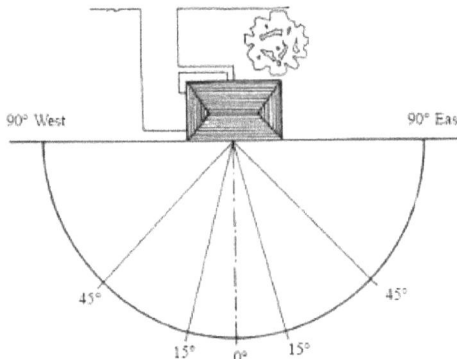

(Figure 20). The optimal angle of the longest house wall with the true south

1.4.7 Active solar power supply method (active)

As mentioned earlier, buildings are able to meet their heat needs from the sun in two ways: 1) Passive solar system (passive), 2) Active solar system. After a complete review of the passive system above, we now turn to a full review of the active system.

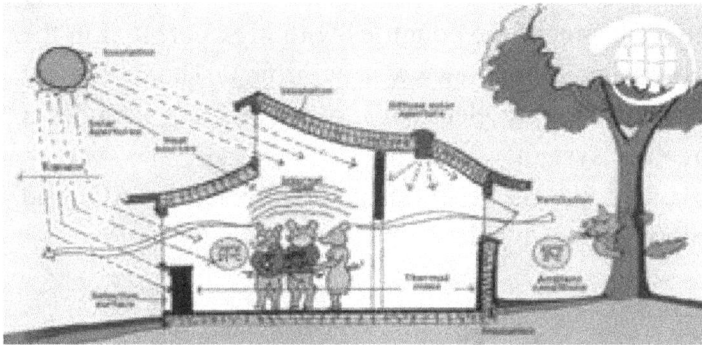

(Figure 21). Use of solar energy in buildings

Passive solar homes can be considered exceptional homes that the main difference between them and ordinary buildings is the way they collect solar thermal energy and their unique and precise design. In most passive solar homes, the rate of air inflow and outflow was zero and as a general result, they need a mechanical air exchange system to achieve the desired indoor air quality (Office for Development and Promotion of National Building Regulations, 2002). The components used in these systems are: 1) Collectors, 2) Thermal energy storage system, 3) Fluid passage channels, 4) Pumps, 5) Piping, 6) Valves, 7) Dampers, 8) Manual or automatic control systems, 9) Auxiliary fuel system and heat tranducers.

1.4.7.1 PV photovoltaic

The intensity of the sun's radiation to the earth varies in different parts of the world, but in general, the closer we get to the equator, the more this radiation increases. Our country, Iran, is located at 25 to 40 degrees north latitude and 44 to 64 degrees east longitude. With about 300 sunny days a year and an average radiation of 4.5 to 5.5 kWh per square meter per day, it is one of the countries with high potential in the field of solar energy (www.irimo.ir/far). The necessary equipment to supply electricity to the building includes a photovoltaic system:

Photovoltaic panels, grid insulators, batteries, AC grid insulators, as shown below.

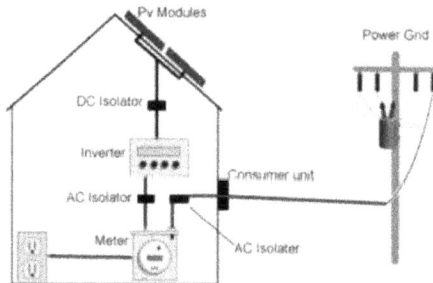

(Figure 22). Schematic of connection of photovoltaic system in the building

The efficiency of photovoltaic panels depends entirely on other parameters such as the level of air purity and the installation angle of the panels. Also, the maintenance of these equipments has a great effect on their lifespan and prevents the reduction of efficiency.

How Solar Cells Work

Silicon can not hold its electrons; thus the cells are made of two layers of silicon; one has an excess electron and the other has an electron deficiency. When light hits the first layer, electrons are released and as they flow toward the layer with fewer electrons, they pass through an electrical circuit and generate electricity. Modules that are installed on a plate and have the correct angle and direction for maximum seasonal and annual collection are called panels or PV grids. Single-module panels can form huge collections of PV networks and generate different DC voltages that can be converted to any desired DC or AC voltage with the help of electronic converters.

Suitable conditions for installation and running of photovoltaic systems

Maximum efficiency of solar collectors is that the panels are perpendicular to the direction of radiation. Reversible panels are still the best option in hot and dry areas but even in these conditions, 10 to 20% of energy is wasted due to reflection. Nowadays, rotating panels can only be considered in hot and dry climates or when the installation of rotating panels is cheaper than their combined installation as building materials. Normally, the best orientation is on the southern aspect.

In addition to direction, the angle of inclination of the panel is also an important factor in the design of solar systems. The angle of inclination is the angle that solar panels make with the horizon and its amount varies from zero to ninety degrees. Due to the deviation of the earth's axis, the angle of the sun's rays changes throughout the year. Therefore, the slope angles of the panels are different in winter and summer. In summer the slope angle is usually between 10 to 15 degrees less than latitude, in winter the slope angle is about 10 to 15 degrees higher than latitude and in spring and autumn is equal to the latitude of the target area. Snowfall is also effective in the amount of the optimal slope angle. In order to prevent the accumulation of snow on the solar panels, the slope angle is usually considered to be about 60 degrees. The table below provides appropriate slope values of the panels for different latitudes in details.

(Table 2) Suitable amount of slope panels for different latitudes

Appropriate slope of the plates	Latitude of the area (degree)
15	0-15
Equal to latitude	15-25
latitude 5+	25-30
latitude 10+	30-35
latitude 15+	35-40
latitude 20+	+40

Integrated Photovoltaics (designed for the building)

Combined photovoltaics can replace ceilings, facades, shutter walls, glass work or special elements such as canopies or sunshades. Here are some key benefits to using them include:

Eliminate the cost of transporting electricity to the building, which can save more than 50% of the cost of electricity.

Eliminate energy loss along the transportation route, which can save up to 25% of electricity costs.

Avoid wasting open space, which is a very important element for picking up photovoltaic panels;

Avoid wasting money on part of the building cover by using photovoltaic panels; avoid creating a supporting structure for photovoltaics;

The beauty of using a new material to cover a building;

Profit from generating all or at least most of the electricity required by using environmentally friendly tools.

Cover the roof with photovoltaics

The best condition is on a sloping roof and if the roof is flat, it can be made with a structure. But it must be such that the integrity of the building is not disturbed and the slope is toward the south.

Cover the facades with photovoltaics

By covering the east and west facades with panels, more than 60% of the energy produced in the south can be produced. They should not only be shaded, but also have air flowing behind them.

Work with glass and photovoltaics

Installing photovoltaics instead of glass is generally suitable in places where there is no need for light or there is no problem in terms of vision and perspective. Today, transparent photovoltaic cells are being developed that can produce both cool indoor air and electricity by covering the bodies with these panels exactly what most buildings in the tropical regions need to.

Create shadows with photovoltaics

Creating shadows with photovoltaics is a very good method and application for these panels.

1.4.7.2 Solar Thermal

A large amount of the energy consumed in a building is related to heating sanitary water for washing dishes, bathrooms and toilets. This amount of energy can be generated and stored by the relevant equipment during the hours of the day when there is solar radiation, so that hot water can be accessed during the whole day.

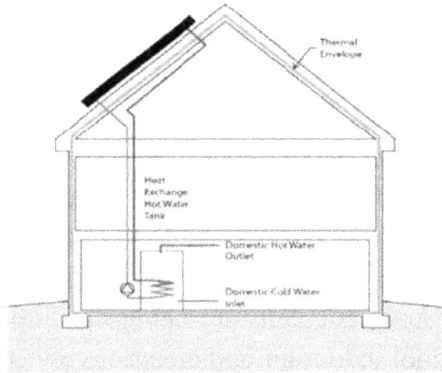

(Figure 23). Schematic of connecting the solar heating system in the building

The components of a solar heating system are:
Solar collector plates
Hot water storage tank
Water pump (in pressurized and non-gravity models)
Water piping between equipment

1.4.8 Types of Passive Solar Cooling Methods in Building

Natural cooling techniques keep the house cool without using any energy in the summer. Shade is one of the most practical and important items in passive solar houses because the same structure collects sunlight in winter. Thermal mass and building materials are as effective in cooling as they are in heating. They store heat in winter and are used to cool the house in summer. Passive cooling is also a method of using windows that transfer less heat to the house in the summer by creating shade.

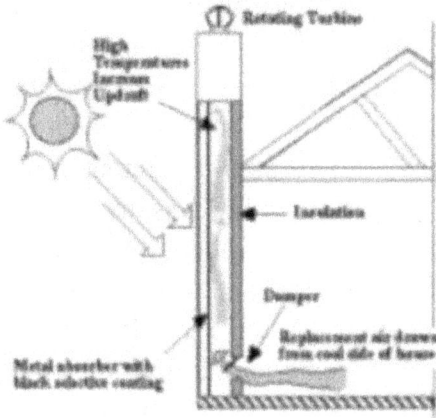

(Figure 24). Passive solar cooling methods in building

The effect of color on the amount of solar energy absorbed on a surface

Before sunrise, the temperature changes of the external surfaces of the building walls in all directions are almost synchronous with the changes of the outdoor air temperature. In this case, only the roof temperature is several degrees lower than the outside air temperature because it loses most of its heat by emitting a long-wavelength beam into the sky. But after sunrise when the sun's rays are reflected directly from other surfaces reflected from the sky. The temperature of the outer surfaces of the walls increases in proportion to the radiation they gain and absorb. If the surface is light in color, the amount of heat absorbed by that surface due to sunlight is low and the temperature of the air around that surface has a greater effect on the heat produced on that surface. But when the color of an outer surface is dark, the effect of sunlight on the heat produced in it is much greater than the effect of the temperature of the air around that surface (Watson, 2013).

In an experiment, a light 2 cm wall was once placed in gray and once in white in different geographical directions. They came to the conclusion that in terms of receiving solar energy, there is a great, mutal correlation between the color and the orientation of the surfaces. If the color of the wall is gray, there is a difference of up to about 3 degrees between temperatures of the walls surface in different directions. But if it is white, the difference will be less than 2 degrees. This result shows that the discussion about the orientation of the building without considering the color of the exterior surfaces is meaningless. Another result is that by using different colors on the exterior surfaces of the walls of a building, the effects of the sun's radiant heat in its interior spaces can be controlled. The difference in wall temperature level in these two experiments is due to the properties of absorption and repulsion of solar energy and different colors. Light colors may reflect up to 90% of solar energy, while dark colors may reflect only 15% or less (Watson, 2013). As a result, considering that the amount of sunlight on the southern surface and the angles close to, it is much higher in winter than in summer. In order to get the maximum solar energy, dark color is more suitable for the southern front than light one. Especially if the south walls have a horizontal canopy so that there is not the same little summer light on the wall (www.newsfaspoort.ir).

Cool Roofs
Roofs with a white thermoplastic layer naturally reflect light and have the highest amount of reflection and propagation on the roof. For example, a roof made of white thermoplastic can reflect 80% (or more) of sunlight and emit at least 70% of the heat absorbed by the roof. A pitched roof reflects only 6 to 26% of sunlight.

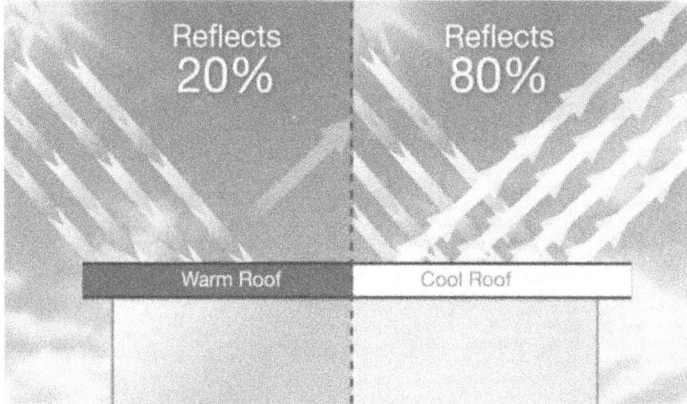

(Figure 25) Reflection of sunlight by the bright color of the roof

The highest SRI rates and the coolest roofs are stainless steel roofs, in the case of normal winds, they are only a few degrees warmer than the ambient temperature. The SRI range for them is from 100 to 115. Some of them also have hydrophobic properties; so, remain very clean and they retain their original SRI even in contaminated environments.

(Figure 26). Reflection of sunlight by the bright color of the roof

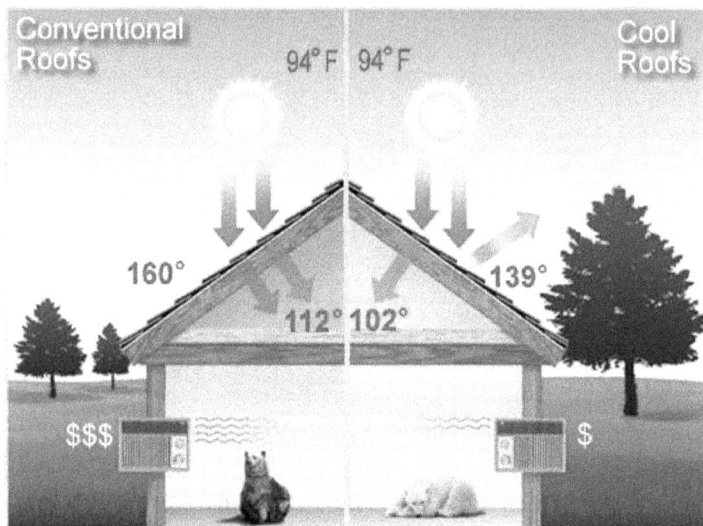

(Figure 27). Impact of cool roof on home space

1.4.9 The effect of sunlight on the internal temperature of the building

Heat transfer can take the form of conduction, convection, radiation, and evaporation, which are briefly described (Kasmaei, 2003).

Condution

Passing from one molecule to the next, heat can pass within the objects or from one object to another in contact with it. This type of heat transfer is called conduction. Usually, the denser the material of a wall, the faster heat passes through it in the form of condution.

Convection

Air can transfer heat from hot surfaces to cold surfaces. This type of heat transfer is called convection. Air can cause heat transfer (convection) if there is enough space for; otherwise, (such as the air inside the ionolite) not only is it not able to transfer heat but it is also a good thermal insulator.

Radiation

Heat, like light, is transmitted electromagnetically, called radiation. By radiating long-wavelength electromagnetic waves, heat is transferred from a hot to a cold object without affecting the air between the two surfaces.

Evaporation

Evaporation is the change of temperature and heat transfer due to the change of liquid to steam. This deformation causes heat dissipation. In double-walled walls, heat is conveyed and radiated from one side of the wall to the other, then transmitted to the inner surface of the wall and causes it to heat up. Indoor surfaces, after heating, transfer their heat in the form of convection and radiation to indoor air and other surfaces.

1.4.10 The effect of sunlight on building components
Sunlight on all types of walls

Increasing the outside air temperature causes the outer surface of the side walls of the building to heat up. This effect is the same in all aspects of the building and the direction of the walls has no effect on the amount of heat gained in this case. In addition, the sun warms the wall, in this case, the direction of the walls is quite effective in gaining heat. For this reason, different temperatures are created in the walls and the whole building under the influence of sunlight (Kasmaei, 2003).

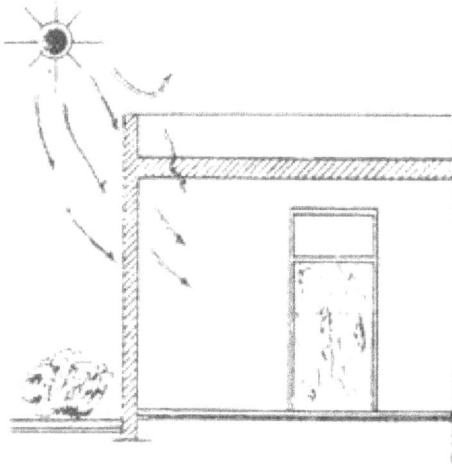

(Figure 28). Sunshine on the wall

It can be imagined that the wall is composed of several layers. Due to the transfer of heat to each of the layers, their temperature increases and the amount of heat that causes this temperature rise, is stored in that layer. Then it is transferred to the next layer. Therefore, each layer of the wall receives less heat than the previous layer. As a result, its temperature decreases compared to the previous layer. As heat is stored inside the wall, less heat is transferred to its inner surface and its temperature is lower than the outside surface temperature. After the temperature of the outer surface of the wall reaches its maximum, it begins to cool down due to the decrease in the intensity of the sun and the cooling of the air. The flow of air inside is reversed. At this stage, first the heat stored in the wall moves inwards and outwards in two directions but then it just moves out. As a result, the wall layers gradually lose their heat and cool down. In this way, the walls of the building go through a period of warming and cooling during the day. Of course, the fluctuation range of this period of heating and cooling is not the same for internal and external surfaces. Temperature fluctuations on internal surfaces are always less than on external surfaces. The internal surfaces reach their maximum and minimum temperatures some time after external surfaces as well. Assuming that the outdoor air conditions are constant, the maximum and minimum temperatures of the indoor surfaces and the ratio of the temperature fluctuation of the indoor to the outdoor surfaces depend on the capacity and thermal resistance of the exterior wall materials. The higher the capacity and thermal resistance of a wall, the lower the temperature fluctuation of the interior surface and the time to reach the minimum and maximum temperature of indoor surfaces is more delayed than outdoor air. Reducing the temperature fluctuation of the interior surfaces of a building compared to its exterior surfaces is proportional to the thermal resistance of its wall materials. However, the delay in reaching the maximum and minimum temperature of internal surfaces compared to external surfaces depends on the thermal capacity of wall materials (Idem).

The effect of sunlight on the window
When the sun shines on transparent surfaces, its beam is divided into three parts. Part of it reflects that this beam has no thermal effect on the transparent object. The other part is absorbed by the glass and then it is transferred to the surroundings as thermal energy. And the third part passes directly through the glass or transparent object and affects the space behind it. The amount of beam that passes directly through the glass depends on the angle at which the beam strikes the glass surface. If this angle is greater than 45 degrees, less light will pass through the glass. When the angle of impact is greater than 60 degrees, the amount of light passing through the glass is greatly reduced while the amount of reflected beam increases. The amount of energy absorbed in the glass body is not related to the angle at which the beam strikes the glass surface.

The effect of window direction

The effect of window orientation on room temperature largely depends on the natural ventilation condition of the room and the condition of the window canopy. In one experiment, indoor air temperature change curves were drawn in samples with no air flow and their window had no canopies. Before sunrise, the indoor temperature of all samples was the same but during the day, this temperature changes with respect to the orientation of the specimens. Immediately after sunrise, the indoor air temperature of the windows facing east increases by 13 ° C within 4 hours whereas the increase in outside air temperature during this period has been only 5 degrees. The temperature rise of the samples to the west is very low until noon. But in the afternoon, after receiving the sun's rays, its indoor temperature rises 11 degrees above the outside temperature. At the same time, the indoor air temperature test (July) of the samples facing north and south is the same. The difference is that the increase in indoor air temperature of the samples to the south is more at noon. The maximum increase was 3 degrees above the outside air temperature (Idem). The same previous experiment was performed with ventilation and draught. In this case, the difference between the indoor air temperatures of the samples compared to the different directions of the windows is very small and indoor and outdoor air temperature changes are close to each other. The effect of window canopies and natural ventilation in determining the indoor air temperature of a building is much greater than the direction of the windows. In a room where the windows have effective canopies and the air is flowing in it, the placement of the windows has no effect on the indoor air temperature.

Sunlight in Double Glazing - Reflecting Sunlight
In this type of glass, a glass surface is covered with a coating that reflects light and heat.
If we look at them from the outside in daylight, they are like a mirror and if we look at them from inside the building, the glass will be completely clear. At night, this will be the opposite.

This glass significantly reduces the heat caused by sunlight by reflecting sunlight. As a result, it saves on the cost of construction, operation and maintenance of air conditioners and conversion. These glasses reflect more of the different spectrums of light and they are very effective in controlling the entry and exit of light and energy. In choosing reflex glass, it should pay attention to its limitations, including being a mirror at night from the inside. This glass has many fans due to the variety of colors and visual beauty. Reflex double glazing is typically made of two layers of glass, one clear glass and the other reflex glass. According to the calculations, the amount of light and heat transmitted in the double-glazed reflex glass is 61% and 63%, respectively.

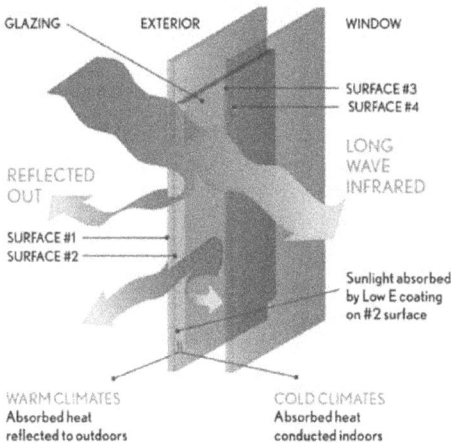

(Figure 29). Reflection of sunlight by window glass

Low-e Glasses

These glasses reflect infrared heat rays but transmit visible light. In the tropics, choosing this type of glass prevents the loss of thermal energy inside the building to the outside. The performance of low-emission glass in controlling the conduction and radiative heat transfer in the skylights of a building is measured by two indicators, U-Value and Shading Coefficient. U-Value or heat transfer coefficient indicates the amount and speed of heat transfer through the conduction between the building and the surrounding environment. Shading coeeficient indicates the amount of passing solar radiation energy. As mentioned, the heat transfer coefficient of Low-E double glazing is reduced by half by ordinary double glazing and the range of Shading Coefficient in low emission glass is very wide. Low-E glasses themselves have types with high and low Shading Coefficient. Glass with a low shading coefficient absorbs a large amount of the sun's heat energy and they help to heat the building. This type of glass is especially suitable for cold areas or low sun views. On the other hand, V-Cool glasses with a high Shading Coefficient prevent the sun's heat from entering the building and they reduce significantly the cost of building cooling in tropical or sunny areas.

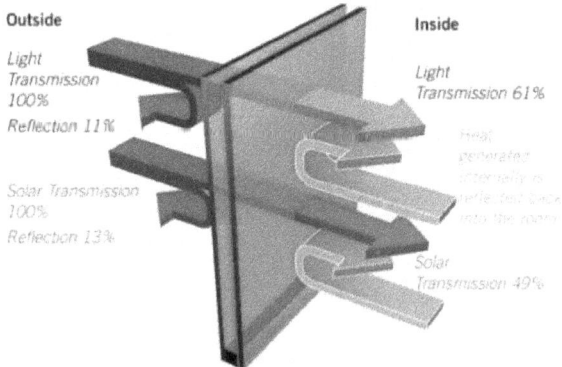

How it Works:
Example: *4mm Solar Neutral Tough / Gas / 4mm Elite Tough*

Outside

Inside

Light Transmission 100%
Reflection 11%

Light Transmission 61%

Solar Transmission 100%
Reflection 13%

Solar Transmission 49%

(Figure 30). Low-e glasses

Control Glass
Control glass, such as "colored/stained glass", is widely used to reduce the sun's rays, heat and light, while being delicate. In these glasses, the amount of sunlight can be reduced by up to 75% compared to ordinary glass. Obviously, the best way to use colored glass is to remove it during the winter to make the most of the sunlight and heat. For example, stained

glass can replace a regular double-glazed glass used in winter to absorb the sun during the summer (Cook, 2018).

(Figure 31). Control Glass

Effects of doors and windows on the thermal comfort of the building and executive solutions.

Much air intake can be reduced by using inlet porches, storm windows, windcatchers and opening orientations. Also, with the correct design of details and selection of seamless doors and windows, air infiltration can be greatly reduced (Cook, 1997).

The doors and windows of every house are used to waste heat in winter in two ways. Doors and windows are weaker in terms of insulation than opaque walls. As a result, they reduce the total thermal resistance of the building shell. As the number of doors and windows increases, the amount of air infiltration rise as well (Idem).

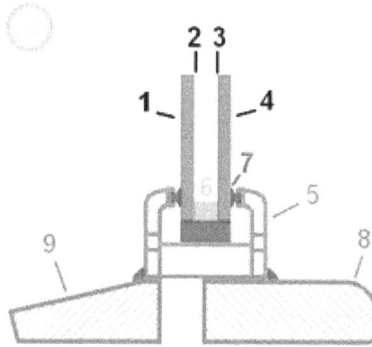

(Figure 32). Cut from an insulated window (Insulated Glass Unit: IGU)

Insulated windows are ones that are commonly known as double glazed or double glazed windows or triple glazed windows. Standard insulated windows, while reducing energy consumption by 25 to 40 percent, provide thermal comfort to building occupants.

1.4.11 Types of Skylights

The skylight or cailing window transmits the light. These skylights form all or part of the roof space of the building for lighting. Types of skylights include top roof windows, single skylights, tubular lighting devices, sloping glazing, and standard skylights.

(Figure 33). Optical tube

(Figure 34). Bubble roof skylights and ordinary skylights

1.4.12 The Effect of Canopy

When shadows form on the outer surface of the glass, very little of the sun's heat energy is transferred to the space behind the glass because heat transfer in this case is in the form of (conduction) and (radiation) and heat transfer rarely passes through glass in the form of conduction as well as the transparent objects also do not transmit long-wavelength rays. But when we use roller shutters to prevent direct sunlight from penetrating inside, the direct sunlight passes through the glass and affects the roller shutters by their heat load. After heating, the roller shutters transfer their heat around in the form of long-wavelength waves and this heat is only transferred to the interior because it can not pass through the glass finally causing the space to heat up

(Kasmaei, 2003). The results of experiments conducted in this field show that outdoor canopies reduce the thermal effect of sunlight inside the room by up to 90% and interior canpies by only 20 to 25%. Effect of window orientation inside unventilated specimens; but with internal canopies. If roller canopies are installed on the outside of the windows, the difference in indoor air temperature of the above models will be less.

Table 3. The effect of the type of canopy and window on its indoor air temperature. Source (Idem)

North-south average	East-West average	South	North	West	East	Ventilation condition	Canopy color	Canopy status
2/5 0/45 2/95	8/6 0/4 5/85	2/5 0/4 5/7	2/5 -0/4 2/2	11/4 0/9 7/7	5/9 0/0 4/0	- window and closed opening part - Open window - close window - Open window opening		No canopy
0/25 0/28	0/95 0/25	0/0 0/65	0/2 0/65	0/5 0/25	0/0 0/2	- open window closed - Open window opening	dark	Exterior canopy
-0/5 -0/1	0/0 0/2	-0/1 0/0	-0/3 0/2	0/3 0/2	0/3 0/9	- closed window opening - open window	light	
2/3 1/05	5/3 2/1	2/0 1/5	9/6 0/6	8/0 6/5	2/6 9/7	- closed window opening - closed window opening	Dark light	Interior canopy

Types of Canopies

Canopies can have a variety of effects, such as controlling direct sunlight (continuously or in person), controlling light, scenery, and natural ventilation. The importance of these effects depends on the location and type of building. For example, in a residential house, direct sunlight may be necessary to penetrate inside in winter and unnecessary in summer. But in a classroom, direct sunlight inside can be uncomfortable in all seasons. On the other hand, in cold regions, the main goal is to use direct sunlight and natural heat of the sun as much as possible. But in hot areas, direct sunlight should be avoided as much as possible (Idem).

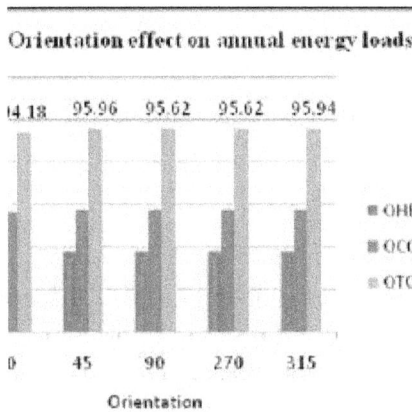

(Figure 35). Effect of window orientation on indoor temperature of unventilated specimens, but with dark inner canopies. Dark color. Source (Kasmaei, 2003)

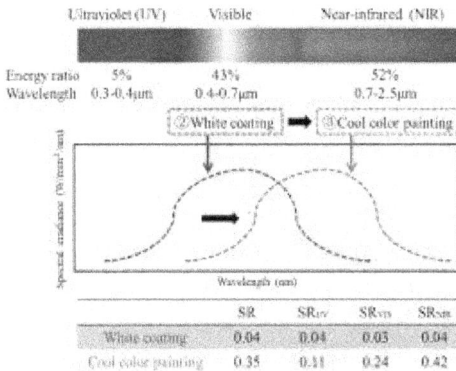

(Figure 36). Effect of window orientation on indoor temperature of unventilated specimens, but with outer canopies. Dark color. ource (Kasmaei, 2003)

Movable Canopies

The geometric shape of the horizontal or vertical movable canopy does not affect their efficiency in terms of providing shade and preventing sunlight because these canopies can be changed as needed. On the other hand, the efficiency of these canopies is different and color and location of their installation depends on the window and the conditions of natural ventilation in the building. Surveys of various types of movable canopies show that;

1) Exterior canopies are much more efficient than indoor awnings if movable ones are installed on the outside of the window. Only 5% of the solar energy radiated to the window is transmitted inside.

2) The darker the color of these canopies, the greater the difference between the performance of the indoor and outdoor types of these canopies.

3) The darker the color of the outer canopies, the more efficient they are.

4) The lighter the color of the inner canopies, the more efficient they are.

5) By using efficient canopies such as exterior wooden roller windows, it is possible to prevent the penetration of more than 90% of solar thermal energy due to sunlight on the window inside.

6) Dark interior movable canopies (roller shutter) transfer 70 to 80% of solar energy to the window.

By darkening the color of the canopies and closing the windows, their efficiency can be increased but if the windows are open, the effect of the color of the canopies largely depends on their position relative to the direction of the wind. For example, if the wind blows from the west in the afternoon and the windows are open, the dark canopies of the western windows allow the heated air to enter the room due to contact with them and heat the space. In this particular case, if the thermal capacity of the canopy materials is high, their thermal effect will remain for a long time after sunset. When windows are open, if the dark-colored canopies on the outside are back to the wind, they have less effect on the indoor air heating because the air that comes in contact with them, gets away from the building.

Fixed Canopies

The most effective canopies for south, southeast, southwest, east and west windows are frame-shaped canopies. Especially if the vertical parts of these frames are located at an angle of 45 degrees to the south, they are very suitable for creating a useful shadow. For these two directions, horizontal canopies are more suitable than vertical ones. In fact, vertical awnings, even at very high altitudes, not only provide very little shade to the window in summer, but also prevent direct sunlight in winter.

Since these awnings are fixed in all seasons of the year, their efficiency in terms of creating effective shade on the windows depends on the position of the building and the daily and annual changes in the position of the sun and regarding the efficiency of fixed canopies, it is important to pay attention to geographical directions (Kasmaei, 2003).

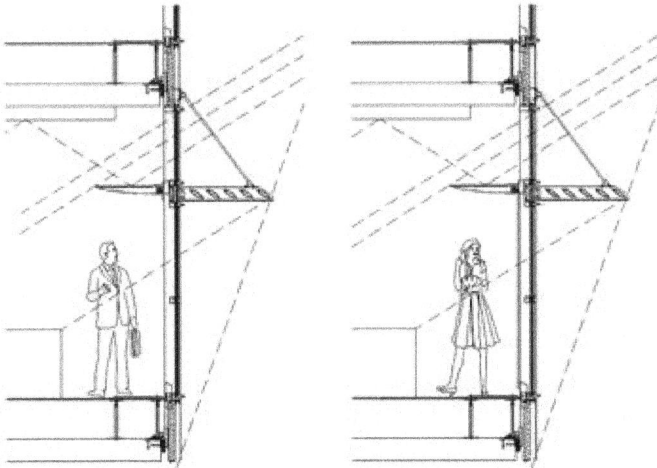

(Figure 37). Fixed canopies

Aluminum Lor (sunshiny) in Building Facade

Lor, as an element in the facade, make it possible to control the entry of natural light into the building in glass facades.

(Figure 38). Lor Nama (Sunshiny)

This type of shading is always one of the available options. Although they protect against glare, they distort outward vision based on how open or closed the roller is. This type of shading, according to their adjustment, directs the light to the depth of the space and produces a homogeneous light.

(Figure 39). Function of Lor Nama (sunshiny)

Light shelf for redirection of sunlight

These shades are often used in warm climates with sunny skies and they are installed next to the vertical windows that are above the level of visibility. They are divided into two types, internal and external. Of course, the combination of light niches inside and outside the building helps to homogenize the light that reflects into the interior and the depth of its penetration.

(Figure 40). Light shelf for redirection of sunlight

Glazing with reflecting profiles

Reflective glass is used in all climates and can be installed in vertical and ceiling type. If this type of shade, used in its ceiling type, protects less from glare, can not be seen outside and transmits light less to the depth of space. If it is used vertically, depending on the installation location; it creates homogeneous light, does not require light tracking and is fully accessible.

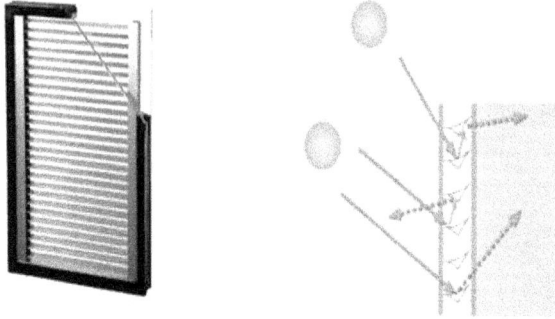

(Figure 41). Horizontal reflector of sunlight or light niche

Skylight with laser cut panels

This type of shader is used in warm climates with sunny skies and is usually used for roofing. Depending on the design, this type of shader can control glare and transmit homogeneous and uniform light to the depth of space. It does not need to track the sun and creates a beautiful and pleasant atmosphere.

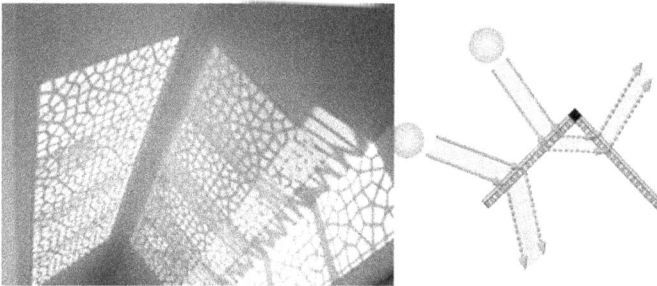

(Figure 42). Skylight with laser cut panels

Natural Canopies

Trees can be an important factor in creating private spaces and at the same time greatly reduce the intensity of sunlight. The sticky surface and under the leaves of trees absorb airborne dust and keep it clean. If the trees are planted densely and compactly, they will have a great effect on controlling and reducing noise. Moreover, the most important feature of trees in architecture is their effect on the thermal condition of the building. Evergreen trees such as pine and cypress reduce wind pressure on the building, thereby preventing heat loss in the winter. In summer, the surface of plants and leaves of trees absorbs the sun's rays, and the evaporation that takes place on these surfaces causes the air to cool. Most importantly, the trees provide the appropriate shade with their climate Kasmaei, 2003).

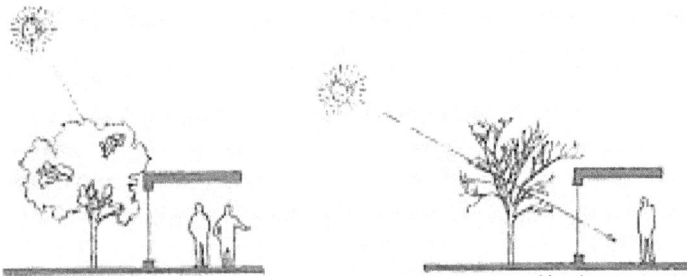

(Figure 43). Natural canopies

Examining this figure, it can be concluded that although the leaves of the trees go a little earlier than when shade is needed, the amount and total shade that each group of these trees creates is perfectly proportional to the climate of their location. Even the change in the condition of the trees throughout the year is commensurate with human needs. This means that the leaves of the trees grow later in the colder winters and fall later in the warmer and longer summers. They are especially valuable when trees are planted near the building because in this case, their leaves fall in winter and they have leaves again in summer, and this is the main principle and design of canopies (vine tree and ivy).

Choosing the right type and location of the tree for each area is very important. In this case, you should always pay attention to two factors: the shape and function of the tree in summer and winter, as well as the shape of the shade that the tree creates at different times of the year. The exact location of the tree should be determined according to the type and shape of shade that the tree creates in summer. Trees should be planted in an area of the building that casts a shadow on the surface during hot hours and when the sun is shining at its maximum energy. Due to the low angle of radiation in the morning and evening, trees are the best type of canopies for windows and walls in the east-west and southeast-southwest of short buildings. During these hours, the horizontal rays of sunlight create elongated, tall shadows of trees that can effectively cover these walls (Idem).

Without a tree, it is very difficult to cast shadows on these walls. At noon, the angle of the sun is high and the trees can not cast a proper shadow on the building, but in these hours, using a very simple horizontal canopy, provides a suitable shade for the windows of the south walls.

1.5 Vision of the use of renewable energy in the world

Exploring the perspective of renewable energy by countries around the world will be a new way to examine the penetration of these systems in future decades. In this regard, companies in the field of energy in the world in numerous reports have provided statistics and forecasts for the use of renewable energy in the form of scenarios which studying number of these scenarios for a better understanding of the subject is described below.

The presented scenarios are introduced in three sections: pessimistic, optimistic and intermediate. The pessimistic scenario that envisions the worst future for renewable energy has been presented by oil companies and firms, the CIA and the International Energy Agency. In pessimistic scenarios, the maximum share of renewable energy in the future until 2050 is only 15 to 20 percent.

In other scenarios, which are presented in the form of intermediate scenarios and the World Climate Change Commission is in favor of this scenario, the share of renewable energy use is considered to be 25 to 40 percent.

The third scenario is called the optimistic scenario. Global Energy Assessment (GEA) reports predict a 95% share of renewable energy by 2050. Another scenario is called sustainable development by the World Energy Agency and the sample roadmap proposed by the German Association for Climate Change. The share of renewable energy in the world for 2050 is projected at 35% and 50%, respectively.

In the sustainable growth scenario presented by Shell Oil Company in 2010, the share of renewable energy by 2050 is estimated at 50%.

Given these scenarios, even if we abandon the optimistic scenario and use the intermediate scenarios as a basis, the importance of the place of use of renewable energy in the future becomes more and more clear (Farazmand, 2014).

These reports indicate the expansion of the use of renewable energy in the coming years. There are many motivating factors in this regard that encourage governments and people to use new energy sources. Some of these factors are: job creation, industrial development, risk reduction, price fluctuations in fossil fuels, climate change, environmental sustainability and nuclear accidents and waste.

The mentioned factors, especially the environmental pollution of fossil fuels and the depletion of the sources of these fuels have doubled the motivation to use new energies to the extent that in the planning of some countries, the reception of these countries has surpassed the presented perspectives. For example, in 2000, the International Energy Agency planned to install 34 GW of wind turbines in the world by 2010. However, in 2010 the capacity of wind turbines in the world increased from 34 GW to 200 GW, which is about 6 times (Shamandi and Rasooli, 2016).

1.5.1 Advantages of using Zero Energy Buildings

1- Residents of the buildings will be safe from rising fossil fuel pricesand rising prices will not affect the family economy.

2- Reducing the production of polluting gases by residential houses, due to the major role of environmental pollution by residential houses.

3- High reliability, for example, photovoltaic systems have a 25-year warranty and rarely do they experience problems due to climate change.

4- Reduce the net monthly cost of living

1.5.2 Major disadvantages of Zero Energy Buildings

1- Implementation of this system in buildings requires high initial costs and training of specialized personnel.

2- Lack of technical knowledge, skills and experience required in the design and construction of zero energy buildings.

3- Photovoltaic cell technology has reduced prices by about 17%. This will also reduce the cost of investing in solar-based power generation systems.

4- Decreased ability to sell such buildings due to initial costs and the need for tough competition in sales.

5- Solar energy absorbed through the shell of the building is most efficient only in its southern part, and in other directions its efficieny will be more reduced due to the shadow (S.C M Hui zero Energy Building, 2008).

1.5.3 Zero Energy Consumption position in buildings

About one-third of the world's energy is used in construction. According to the 2013 energy balance sheet, the amount of energy consumption in the construction sector in Iran is 4 to 5 times higher than the standard of European countries (Energy Balance - Developments in the Energy Sector in Iran, 2013). This amount is very significant and increases the motivation to invest in reducing energy consumption in this sector because in addition to limited sources of energy supply, it can be concluded that about one third of environmental pollution and greenhouse gases are produced by the building sector.

An examination of the energy flow chart, published in 2013 by the Electricity and Energy Affairs Planning Office of the Ministry of Energy, shows that the total energy consumption in the country in various fields is equal to 1229 million barrels of crude oil. Accordingly, 440.7 million barrels of crude oil to the share of domestic and commercial sector, which is equivalent to 35% of total energy consumption (Energy Balance - Developments in the Energy Sector in Iran, 2013).

According to this statistic, the share of renewable energy in the supply of energy required by the domestic and commercial sectors is equal to 17.1 million barrels of crude oil, which is equivalent to 8.3% of the total energy required by the domestic and commercial sector, which is very low compared to other countries.

The domestic sector is the largest consumer of kerosene and liquid natural gas in the country and about 32% and 11% of the total energy consumption of these two products in the country, respectively, belongs to this sector.

According to studies, energy consumption in the building sector of Iran and European countries per square meter is equivalent to 80 and 6 cubic meters of natural gas, respectively. In other words, the energy consumption of this sector in Iran is almost 4 to 5 times higher than the standards of European countries.

According to studies, the simultaneous use of photovoltaic and solar heating systems is the most attractive and suitable system for energy extraction for zero energy buildings (Good, 2015). Energy from renewable sources is used for air conditioning, lighting and electrical appliances in the building.

Therefore, the main goal in designing buildings with zero energy is to have zero difference in the amount of energy produced to the energy consumed in the building during a period of one year.

(1) Enet = Eexported – Edelivered

Enet: net energy, which balances zero kilowatt hours for zero-energy buildings throughout the year.

Eexported: energy generated by renewable systems in the building.

Edelivered: energy consumed in different parts of the building.

1.5.4 Energy Balance Line

In any building with zero energy building design (ZEB), the amount of energy consumption in the building and the amount of energy production from renewable sources is always important. In the case of various buildings, the energy balance line is used according to the construction conditions and equipment. To understand the energy balance line, you need to pay attention to figure.

feed-in energy
[*export:* kWh, CO$_2$, etc.]

net zero balance line

energy
supply

starting
point

delivered energy
[*import:* kWh, CO$_2$, etc.]

energy efficiency

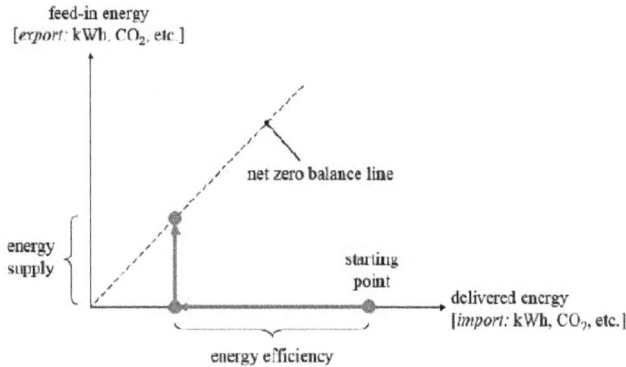

(Figure 44). Energy balance line with zero energy consumption

This diagram shows the energy produced from renewable resources in the building and on the other hand the energy consumed in the building.

Balance graph of zero energy that is linearly marked, indicates the proportion of production sources to consumption. There are two ways to achieve the power building:

1- Increase energy production of renewable resources in the building.

2- Reduce consumer energy consumption, which involves reducing their number and intensity or increasing their returns.

1- Increased energy production of renewable resources

In this section, the use of energy production equipment, such as photovoltaic systems or domestic turbine, geothermal energy, etc., is exploited. This section has limitations due to the limited price and size of residential space, in one size and amount, and can not be increased as a desired because it will lead to the non-economic status of the project. Reduction of energy consumption in energy consuming devices, including lighting, cooling, heating systems, etc., which is done by reducing a small number of them or decreasing the intensity of energy consumption of these equipments or their efficiency. Also use high-tech lighting equipment such as energy-saving lamps (LEDs) whose power consumption is significantly different compared to other systems. In addition, the use of high-efficiency heating and cooling equipment can make it possible to achieve a zero energy balance line.

1.5.5 Lighting equipment with technology: Oled lamps

The method of making OLED lamps was first discovered in 1985 by Ching Tang in the laboratory of Kodak Camera Company. By injecting organic materials that can return to nature, he succeeded in making the first OLED lamp between the cathode plates and LED anodes. With the passing the time and the importance of energy saving and recycling of electronic materials, this technology entered the field of industrial production.

Features of OLED lamps include the following:
Unique response time;
High flexibility;
More sharpness and more transparent light;
By using OLED of PHOLED type, electricity is converted to 100% light.
Application of OLED lamps in lighting:
Until a few years ago, OLED screens were mostly decorative and existed on a small, limited scale but in recent years, they have more and more useful applications. Philips Company recently pursued plans to use these lamps in home lighting design, in which lamps with dimensions of 7.8 x 1.8 inches have a lighting power of 300 lumens per watt. Also, due to

the flexibility of these lamps in bending and ductility, extraordinary designs can be done using these lamps.

(Figure 45). OLED lamps

Energy generated by renewable systems in the building (Eexported)

To achieve buildings with zero energy consumption with different climatic and architectural conditions, different systems can be used, some of which we mention here:

1- Photovoltaic

2- Solar heating

3- Geothermal energy

4- Wind energy

According to what was said, the solar photovoltaic and heating systems were fully described. Next, geothermal and wind energy are described.

1.5.6 Geothermal Energy

Geothermal energy is the thermal energy contained in the Earth's solid crust. This type of energy is often used to generate geothermal electricity, which refers to the cycle of generating electricity from geothermal energy. The technology used in geothermal power generation projects includes dry steam power plants, liquid steam conversion power plants and dual cycle power plants. Geothermal energy unlike other renewable energies is not limited to seasons, times and special conditions and can be used without interruption. Also, the cost of electricity in geothermal power plants is competitive with electricity generated from other conventional (fossil) power plants and is even cheaper than other types of new energy.

Advantages of using thermal energy to generate electricity: 1) Being clean: this method, like wind and solar power plants, does not require fuel, so fossil fuels are preserved and no smoke enters the air. 2) No problem for the area: it requires less space to build a power plant and does not cause complications such as tunnels, open holes, piles of rubbish or petroleum and oil leaks. 3) Reliability: the power plant can be active throughout the year and because of its location on the fuel source, it has no problems with power outages due to bad weather, natural disasters or political tensions. 4) Renewability and permanence. 5) Flow savings: there will be no cost to import fuel and there will be no worries about rising fuel costs. 6) Help to the growth of developing countries: installing it in remote places can raise the standard and quality of life by bringing electricity. Given the benefits we have listed, geothermal energy contributes to the growth of pollution-free developing countries.

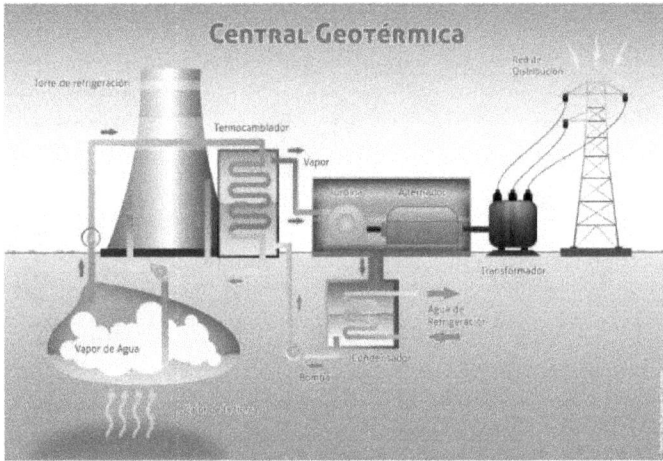

(Figure 46). Geothermal energy

1.5.7 Wind Energy

The use of wind energy in comparison with other renewable energy sources due to reduced electricity generation costs, job creation and lack of environmental pollution in developed countries and many developing countries, has been able to emerge as a new source of electricity supply worldwide.

The use of wind energy in buildings with zero energy approach is obvious because it is possible to install this equipment for buildings. The use of wind energy helps to supply energy to the building.

According to the research, the minimum wind speed for rotating home wind turbines is equal to m/s^2. But in order to be able to use wind energy to generate electricity, the minimum wind speed is estimated at 3.5 m/s and maximum wind turbine power is obtained at speeds above 10 m/s (www.oss.org.uk). The potential of wind energy in the country can be studied and analyzed according to the following map (www.suna.org).

Wind velocity in m/s

<9.0 m/s
<8.5 m/s
<8.0 m/s
<7.5 m/s
<7.0 m/s
<6.5 m/s
<6.0 m/s
<5.5 m/s
<5.0 m/s
<4.5 m/s
<4.0 m/s
<3.0 m/s
<2.0 m/s
<1.0 m/s
– m/s

(Figure 47). Wind speed scattering map in Iran

Based on the wind maps in Iran, most central and eastern regions of the country take advantage from the winds in the speed range of 7 m/s. In other areas, the minimum available speed is 5 m/s and in some areas the wind speed is more than 9 m/s. These numbers indicate a good potential for using wind power to generate household electricity.

Components of a home wind turbine:

1- Wind Turbine
2- Inverter
3- Communication wiring
4- Control system
5- Storage batteries

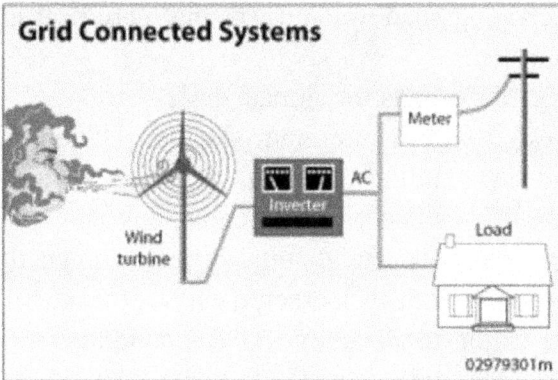

(Figure 48). Schematic of connecting a wind turbine system in a building

To estimate the production capacity of a wind turbine, the following factors must be known:

1- Type of wind turbine

2- Wind speed at the place of installation of wind turbine

3- Angular velocity of the turbine shaft

4- The area where the wind turbine blades sweep

1.6 Climate design (other methods and solutions for designing zero energy building and achieving sustainable architectural goals)

Climate design is a way to reduce the energy cost of a building. Climatic design makes buildings have better comfort conditions. Instead of putting too much pressure on heating and cooling systems, the buildings themselves provide comfort (Kasmaei, 2003).

In winter (or cold weather, when heat is needed) Climate design goals include: Resistance to heat loss and outflow of the building and absorption of more and more solar heat such as sunlight shining from the southern windows. In summer (or hot seasons, when cooling is needed), these goals are reversed, meaning that resistance to the sun's radiant heat is created by creating shadows and wasting more heat inside the building (Idem).

Some suggested methods for climate design

Compact design with a minimum shell in relation to the volume and in buildings with a relatively large area, preferably two floors instead of widening;

In the integration of residential units, the more compactly the units are combined with each other, the lower the ratio of the side surface to the total volume of the building and as a result, the heat loss will be less;

Minimize windows on all fronts except south;

Use double or triple glazed windows and movable thermal insulation on them;

Use adequate and suitable thermal insulation in the outer wall of the building, including floors, walls and ceilings, and avoid creating thermal bridges;

Use two doors at the entrances to prevent heat transducer;

Major placement of buildings under the shade of trees, buildings and other natural features in terms of topography;

Avoid planting evergreen trees on the southern fronts of the building;

Construction of the length of the building along the east-west axis;

Placing the main spaces in the southern parts of the building;

Use glossy surfaces in front of windows to reflect more sunlight inside;

In tropical areas where the "degree/day" of ventilation is higher, insulation is needed only to prevent heat from entering and to maintain a temperature between 25 and 27 inside (Cook, 2018);

If we have few and small double-glazed windows in front of the wind, the air infiltration will be minimized (Idem);

The garage can be placed on the west side of the house, creating a windcatch route or shady yard between you and the house.

A good solution for using solar energy in winter is to place the main walls and windows of the building on its southern front. In most cases, a slight rotation of the building to the southeast is better (Approximately 15 degree). This orientation causes the building to use more sunlight than the afternoon light and the heat absorption by the building will start earlier. The rotation of the building to the southwest makes the house retain the cold air of the morning for a longer time and on the other hand retains the heat of the afternoon until sunset. Doors and windows should be installed in a direction so that summer breezes can enter easily:

The breeze of the air outside the building due to the force of "pressure-suction" causes the movement of air inside the house. Positive pressure is created on the windward side and negative pressure (suction) is created on the other side of the building. For natural ventilation, the openings should be placed in walls with different pressures. The maximum volume of air is moved when windows or doors are located in parts of the building facade where there is a pressure difference.

In the case of the plan, the square of the rear wall has more negative pressure (suction) and as a result will be a more suitable place for the exit windows.

To moderate the weather all year round - design semi-protected areas outside the building:

Patios and porches and enclosures outside the home contribute to the comfort inside the house and also function as a private space.

The roof and porches protrusion in summer casts a shadow on the walls, openings and the area around the building and it helps keep the outside temperature low. As a result, natural air conditioning will be more effective and the conduction of heat through the walls will be reduced. If the plant is planted in enclosed yards and patios and watered regularly or ponds and fountains are installed in these spaces, this action will cause cooling in the building by evaporating water.

In modern homes, radiators and hot air ducts are often located next to an exterior wall or under a window. This is because the exterior walls and windows are the coldest part of the room, so they need more heat to comfort people in all parts of the room. But if the walls and windows are well insulated, the cold in them will be reduced and heating systems can be placed elsewhere.

Design of rooms or special functions in accordance with the direction of sunlight:

If the building plan is designed in such a way that daily activities are in accordance with the path of the sun, there will be better energy savings. If we divide the indoor spaces into hot and cold parts, the efficiency of the above system will be more effective. By placing warm spaces in the direction of the sun, these spaces can obtain the necessary heat from the sun and reduce the use of mechanical devices to a minimum.

Examples of suitable directions can include east or southeast windows for bedrooms, kitchens and breakfast rooms, in which these rooms can use the morning sun in winter. The south-facing window is suitable for the living room and daily activities. Solar rooms or walls can be on the west side (Figure 19h). The west part is not very suitable for residential rooms in that from spring to autumn these rooms will be very warm in the afternoon.

Warehouses and non-residential parts of the house can act as a retaining space such as thermal insulation.

This will be effective when these spaces are located in the windward part of the building and are properly separated from the residential parts of the building. Also, the less used spaces can be located in the western part of the building to absorb the high heat of the sun in the afternoon (especially in summer).

1.6.1 Thermal Insulation

It is necessary to use thermal insulation at any time or any place where the ambient temperature is lower than the human comfort temperature (or when indoor heating should be provided). Also, when you need to use mechanical devices to cool the building, it is necessary to install thermal insulation to provide comfort in summer. The third thing is that the walls that have thermal insulation will be warmer in winter. These walls will provide better comfort in the house and prevent water vapor transpiration on the walls. If insulation is used with materials that have a high thermal mass, it will affect the time of heat transfer absorbed by the sun into the building, although the use of insulation in this way is not very common (Kasmaei, 2003).

In general, thermal insulation has the following functions: 1) Help to save energy, 2) Heat transfer control, 3) Temperature control, 4) Prevention of frostbite, 5) Burn protection, 6) Fire control.

In the tropics, while the cold season is short, insulation is only to prevent entering warm ambient air inside. In this case, the steam protector must be installed on the outer layer of insulation to prevent steam from entering the building. Thermal insulation can be classified into two general forms: 1) In terms of insulation material, 2) In terms of shape appearance (Idem).

Location of Insulation in the Building

In general, insulation should be installed in any part of the building where there is heat transduce. Although thermal conductivity in relation to the ground is often from the inside to the outside, in the tropics this heat transfer may be from the environment and soil to the building (Kasmaei, 2003).

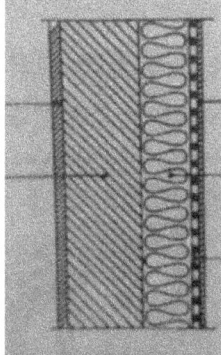

(Figure 49). Insulated wall from the inside with joinery

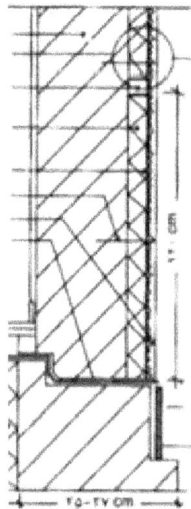

(Figure 50). Insulated wall with facade of cement plaster on thermal insulation

Roof Insulation

Thermal bridges should be minimized in roof design. When thermal insulation is placed on concrete, due to the brightness of the outer surface and the nature of this insulation, the penetration of heat into the concrete during the day is largely prevented, some heat passes through the insulation layer is absorbed into the concrete and raises its temperature slightly. But if thermal insulation is placed under the roof concrete, the concrete layers absorb a lot of heat, enough heat passes through it to increase the temperature of the inner surface. Based on this, the temperature of the roof and the amount of heat transferred into the building through the roof is higher than when the insulation layer is placed on the outside of the roof. The outer surface of the roof in a hot area, even when the roof is completely insulated, must be light.

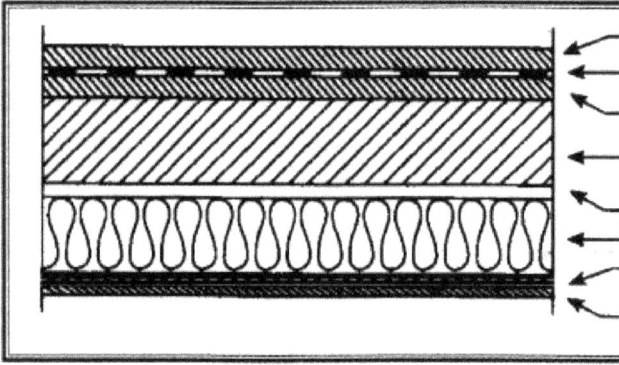

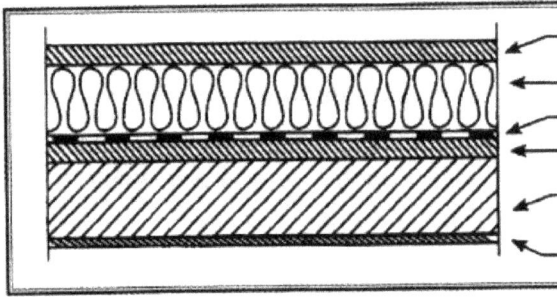

(Figure 51). Right: flat roof with internal insulation on the dry ceiling, Left: flat roof with external insulation on waterproofing

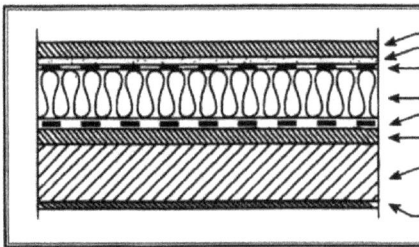

(Figure 52). Right: flat roof with external insulation under waterproofing; Left: thermal insulation around the floor of the building

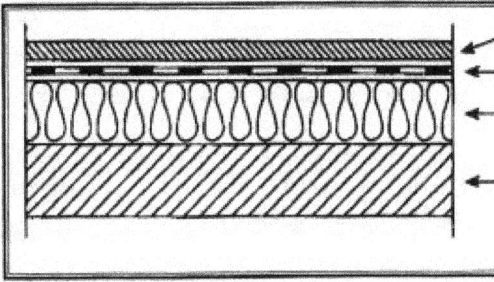

(Figure 53). Right: thermal insulation from inside under the final floor; Left: the state that the water vapor permeability of the outer layers of the wall is high

1.6.2 Climate Design Solutions in Energy Consumption

One of the best ways to save energy is to design the climate of buildings. Sometimes due to improper design of the building, such as placing windows in the wrong place, the amount of opening surface is inappropriate to the climate or the wrong materials in the walls, in times when the outside air is pleasant and good; consequently, the indoor air becomes unsuitable. Buildings should be designed according to different climates and design based on heating in winter and cooling in summer. In climate design, attention to energy consumption and human comfort is best provided. The purpose of this type of architecture is to use passive solar systems, as was common in traditional Iranian architecture. In winter, climate design objectives are: resistance to heat loss and outflow of the building and absorption of as much solar heat as possible, such as the sun's rays shining through the southern windows. In summer, these targets are reversed, that is, resistance to the sun's radiant heat is intended by creating shadows and wasting as much heat as possible inside the building. To achieve these goals, the practical principles of climate design can be used from Table (3) (Mahmoudi and Newey, 2011).

(Table 3). Thermal control methods from hot and cold seasons of the year. Source (Idem)

Energy transfer through					Climate design goal
Evaporation	Radiation	Relocation	Conduction		
	Use the heat of the sun	Reduce outside air flow		Increase heat gain	winter
		Reduce air penetration	Reduce thermal conductivity	Prevent heat loss	
	Decreased solar heat absorption	Reduce air penetration	Reduce thermal conductivity	Prevent heat gain	summer
Use evaporative cooling	Use radiant cooling	Use air ventilation	Use of ground cooling	Increase heat loss	
	The Sun	Atmosphere		Heat sources	Sources
Atmosphere	The Sky	Atmosphere	The Earth	Sources of heat loss	

1.6.3 Roof of Building

The roof of the building has the most radiation and heat gain in summer. Therefore, it should be protected from heat in the summer. The roof also has the highest heat loss in winter. Thus, covering the roof insulation of the building from the inside and placing other components on it is necessary to not absorb heat in summer and not to lose heat in winter. You can also use the pool/pond on the roof of the building. The pond should have a movable cover to be opend during the day and closed at night to absorb solar energy during the day and transfer it to the building at night. The reverse is true in the summer. Also, the shape of the roof of a building, depending on different climates, may be flat, domed or sloping (Mahmoudi and Newey, 2011).

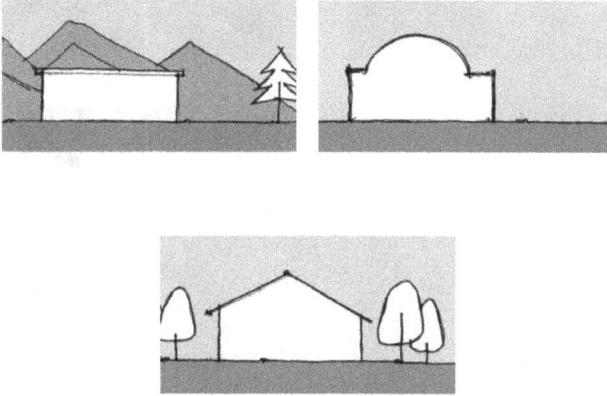

(Figure 54). Roof of buildings in different climates. Source (Idem).

Roof Glass Tiles (absorbing sunlight)
The Swedish company SolTech has invented a kind of glass roof tile. These glass tiles trap the hot air caused by sunlight and it is used to heat water for the home heating system. The tiles are made of ordinary glass and are the same weight as ordinary clay tiles. The tiles are installed on a black nylon canvas with air gaps underneath. The black color absorbs the heat of the sun and the air circulation begins. Hot air heats the water connected to the home heating system by a battery. This process can reduce energy costs and it also be used all year round, even at longer times due to its capacity to store heat in its insulated air layers.

(Figure 55). Roof glass tiles (absorb sunlight)

Architectural Design of Spaces

If the building plan is designed in such a way that daily activities are in line with the path of the sun, it is possible to save a lot of energy. For example, if the indoor spaces are divided into hot and cold spaces, the efficiency of the system will be more effective. By placing the living spaces, kitchen and bedroom in the direction of the sun on the south side of the building, the necessary heat can be obtained from the sun and the use of mechanical installations can be minimized. Placing pre-space in the entrance space of the house is also effective in saving energy by preventing heat transduce with the outside air. This is simply the case applicable with dual door systems. The use of solar greenhouses on the south side of the building is a good factor to provide part of the heating in winter. Observing the appropriate height of the rooms in accordance with the climate of each region also provides air conditioning and living conditions for the residents. It has long been common to build high spaces in desert areas and low spaces in mountainous areas (Idem).

Use of Thermal Mass

The use of heat mass is part of a passive process in which the need for energy is reduced by selecting and placing some high-density building materials such as natural stone, concrete, brick, or even water. These materials with high heat capacity help to create favorable conditions and reduce energy consumption, especially in winter, by slowing down heat exchange through external factors or storing heat during the day and losing energy during the night. In new buildings, pool water in the basement can also be used as a thermal mass. In this way, water is heated by solar energy during the day and releases the absorbed energy during the night and heats the surrounding air. Then, the heated air is transferred to the upper floors of the building according to its convective properties (Mahmoudi and Newey, 2011).

1.6.4 Role of Vegetation in the Building

Trees with evergreen leaves protect from wind, storms, and winter snow if planted north of the building. At the same time, they help lower the temperature around the building by evaporating their leaves and passing wind through them. These trees, with their shade, cause the ground around the building not to retain heat inside themselves and to affect the microclimate of the periphery of the building. In contrast, buildings in which visibility is important are mostly planted with shrubs, bushes and grass. This type of planting reduces the reflected radiation reflected by roads, paths, patios, sand

and water. When watering shrubs and bushes in the morning, it will be created cool air around them. They also cool the surrounding air by evaporation and have a cleansing effect against boiler pollution. Deciduous trees are suitable for planting in the south, east, west and in the yard of the complex. From spring to summer and autumn, the leaves of the trees are a good barrier against sunlight to reach the ground, facades and windows. These plant species lose their leaves in winter, and by shedding the leaves, the sunlight warms the receiving walls and the ground around the building again. This heat causes snow and ice to melt and surface water to evaporate. Broad-leaved trees are the best type of canopy because they protect the building in summer and lose their leaves in winter, allowing the building to benefit from sunlight (Cook, 1997).

Irregular or "rough" surface of the shrub helps prevent sunlight and causes less reflection from even a plant surface such as grass (Cook, 1997).

Dense and short trees as well as tall shrubs in the western part of the building are suitable because they prevent the sun from shining at a low angle in the afternoon. Since the sun does not shine in the east and west of the building as much as in summer; so, evergreen trees can be suitable for shading the western walls. Especially if the direction of winter wind is from the west or northwest, the performance of these trees is doubled (Cook, 1997).

(Figure 56). Tree planting methods and its performance in front of buildings

Evaporation and coolness from a beech tree is equal to the amount of cool air that an air conditioner provides with 20 hours of work per day for 10 ordinary rooms. Therefore, the amount of vegetation should be maximized and wherever possible, man-made spaces should be shaded by trees (Cook, 1997).

Green Roof in Zero Energy Buildings

Green roof is an artificial ecosystem that has many ecological, economic and social benefits in urban environments (Savi, 2015).

Green roof is the vegetation on the roof of buildings. Different names are used for this system. These names are: green roof, roof garden, living roof, ecological rood, and vegetated or vegetative roof. The main difference between a green roof and a roof garden is the depth of the growing medium or soil in them.

(Figure 57). Green roof

Benefits of Green Roofing

Since the green roof is a living vegetation, all the ecosystem services that any other green environment has, can also be considered for it. The various major benefits of green roof are:

Improving runoff management:

Green roof cultivation environment can retain rainwater during rainfall and reduce runoff from building gutters. In many developed countries, runoff or so-called storm water is managed and collected (Carter, 2007).

Reduce the effect of urban heat island:

Research on the impact of green roofs and walls confirms the effectiveness of green roofs and walls in improving the temperature of the city (Alexandri, 2008).

Decreasing air pollution: Green roof plants can reduce air pollution (Getter, 2009; Yang, 2008).

Maintain building energy: The effect of trees and their shade on reducing energy consumption in urban areas has been proven (Akbari, 2001). Green roof also reduces the penetration of heat from the roof to the building in summer with the shade created by plants and their evaporation and transpiration action. Also, in winter, the extra insulation created by the green roof reduces the amount of energy needed to heat the building (Peck, 2003).

Reduce noise pollution: Culture medium with a depth of 12 cm reduces the noise to 40 DB (Peck, 2003).

Increases the life of roof membrane compared to conventional one

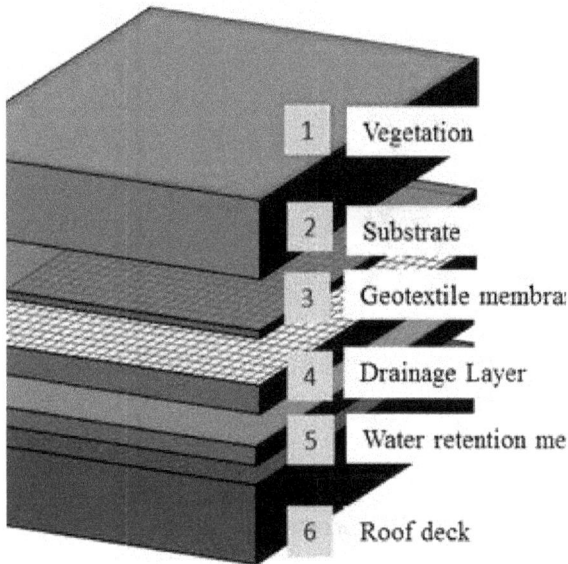

(Figure 58). General schematic view of green roof layers and location of layers

The figure above is an overview of the green roof layers and the location of the layers that are added to them in different sources. For the root barrier layer or material and the vapor control layer, there may be several different locations between the layers.

Green Roofs and their benefits

Green roofs are part of the efforts of city managers to stabilize the urban space and one of the modern solutions to solve urban environmental problems. The main purpose in the past was to use grass to insulate and remove the sealing layer, but today the main purpose is environmental-economic and to improve the management of surface wastewater, health and aesthetics.

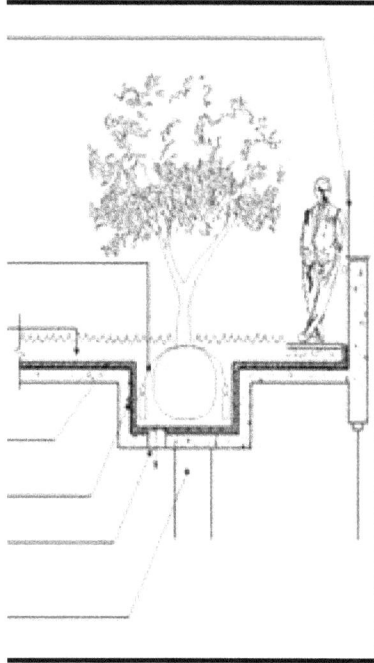

(Figure 59). Green Roof details

The effects of using grass on the roof on the thermal comfort of the building can be mentioned as follows:

Dense grass and vegetation prevent the earth from absorbing the sun's radiation, so that 20 to 30 percent of the sun's energy is reflected and the rest is absorbed by the vegetation. Therefore, the roof surface is shaded and receives much less heat energy than a normal roof.

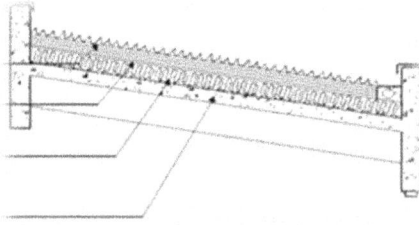

(Figure 60). Green roof vegetation

A well-irrigated short vegetation disperses solar thermal energy during the summer months by approximately 11,350 to 13,630 m2 / kJ via evaporation per day (about 80% of the radiation reaches the ground). Therefore, the net energy reaching the soil is greatly reduced due to the coldness of the environment.

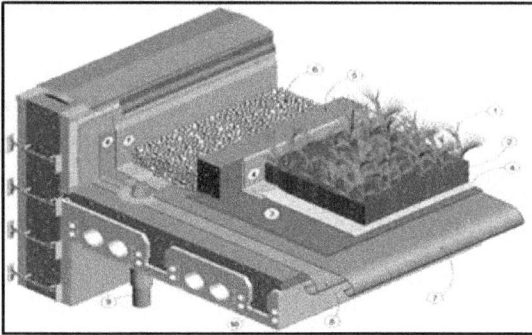

(Figure 61). Short vegetation

The thermal mass of the soil cover reduces temperature changes so that at a depth of 26 cm only 30% of temperature changes are felt. Also, due to the mass of the soil, there is a time delay of about 29.5 hours per meter. Due to the fact that the thermal mass at a depth of 30 to 46 cm reduces the daily temperature changes during winter and summer, the outer surface of the building bears less cold and heat load. As a result, the need for insulation is significantly reduced compared to conventional roofs.

The design and execution of grass-covered roofs allows the use of a mixture of leaves and straw or other similar compounds obtained in the fall as winter insulation. The most important effect of spraying on the roof is to lower the temperature of the roof. In addition to cooling the living space under the roof, spraying creates cool air around the walls.

In relatively good spraying, about 90% of the heat load from the sun on the roof is neutralized. The temperature of the roof exposed to the sun, which is usually around 60-71 centigrade in summer, can be reduced to 26.6 to 37.7 by water spray. In dry areas, the roof temperature can be lower than the surrounding air by spraying during the day.

Green Walls in Buildings

The term "green wall" or "vertical gardens" is a universal term for a living cover system with benefits similar to green roofs. With these gardens, plants grow on and top on the facade of a building. Suitable plants include a wide range of perennial and annual plants and scaffolding of various trees. Vertical gardens, by covering the facade of a building with plants, have a great impact on the surrounding environment. These walls prevent the spread of dust in the air and protect the building from UV rays, rain and wind pressure (Taqawi, 2014).

(Figure 62). View of the green wall

These facades are a type of green wall system in which a variety of short cascading plants are used to cover supporting structures such as modular grid panel systems or cable and wire rope systems. In this system, plants are planted at the base of these structures or in gardens at different levels and it takes several years to reach full growth. Green facades are now divided into two categories:

Multidimensional network system of welded wires;

Network system of stainless steel cables.

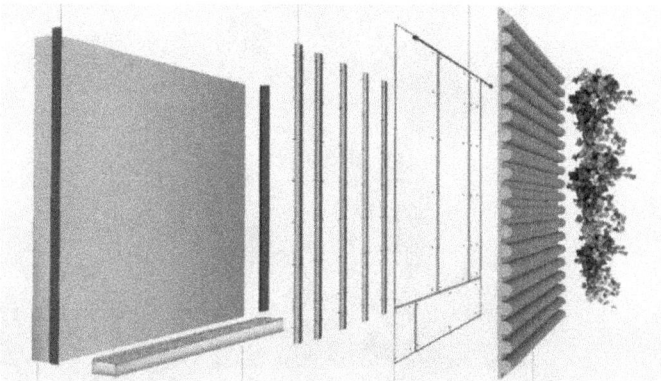

(Figure 63). Green wall components

Green walls have the same function as green roofs and can affect microclimates. These walls are divided into the following two main groups:

Green facades: in this system, the plant moves on the surface of the façade. In this system, a creeping plant or twig, while rooted in the ground, begins to climb the building.

Oxygenator walls: they are divided into active and inactive categories, which are as follows:

Active systems: in this system, the air produced by plants is used in the air conditioning system of the building.

Inactive green wall: it has no role in the movement of air from the roots into the building's ventilation system. Oxygenator walls have an independent structure that is located away from the main facade of the building and at a close distance to it and is restrained by the building. Ivy and creeping plants are very suitable for controlling sunlight on the western facade of the building especially durable varieties can be useful in all seasons. Unlike anti-storm doors and windows, these plants are installed on the building themselves.

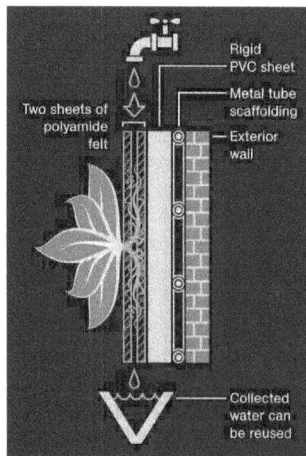

If the plants that grow on the shell of the building are evergreens, in winter these plants will act as insulation between the shell of the building and the environment and prevent heat transfer in the building environment as well. Therefore, for most climatic zones, it is better to plant evergreen plants near the northern, western and eastern parts of the building and use deciduous plants on the south side of the building (Cook, 1997).

1.7 Conclusion of the Subject of Sustainable Energy Consumption

Today, the issue of energy is the concern of all societies because fossil fuels will one day run out. So, using natural and God-given energies is the best way to save energy and fossil fuels for the future. Today, the issue of energy in buildings has received a lot of attention in the world. In order to reach this solution, special attention should be paid to the interaction of architects and other engineers in the building sector at the same time, in order to achieve zero energy buildings. In this process, not only the building must meet all the needs of sustainability as much as possible, but also this possibility will be characterized with the participation of all factors in the construction and implementation and providing opinions of engineers in terms of cost reduction and waste of environmental resources. The entry of these ideas and the time of coordination of strategies is in the early stages of design, are sketches and planning of architectural designers.

Chapter two:

Study of museums in Iran and the world

2.1 Introduction

One of the challenges facing engineers and architects, the confrontation of architectural and engineering considerations, which causes problems in the design of one of them. Therefore, it is very important to pay attention to the considerations of both groups when designing in order to optimize the consumption of the building. Museums usually have exhibits, workshops, offices and warehouses. Some large museums also have shops, restaurants, etc., which will be reviewed in their notes. In the most art museums, galleries and museums of natural history, the display of objects and works is also done in the visiting hall. In museums where the exhibits are separate from the visitors' passages, special conditions may be required because in some cases, the conditions for the protection of objects are significantly different from those for the comfort of individuals.

In museums, working hours are about 8 to 10 hours a day. In museums, the relative humidity and temperature should not change and their amount should remain constant 24 hours a day. In designing museums, the effects of sunlight, especially sun's radiation on objects should be minimized. In winter, it should be prevented distillation of air humidity on external walls and objects close to them.

2.2 Museum

According to the Statute of the International Council of Museums, the museum is a permanent institution with no material purpose, the doors of which are open to all. They work to serve the community and its development. The purpose of museums is to research the relics and evidence left by man and his environment, to collect, preserve and maintain spiritual productivity and to establish a connection between these works, especially to display them for the purpose of study and spiritual productivity (Nafisi, 2001).

According to ICOM's definition of a museum, a museum is a place to collect, maintain, study and investigation, as well as to display cultural or natural blessings in order to educate, research and value these collections and to enjoy them. Today, attention to museums and museum management has the ability to meet all challenges and cultural, social and historical needs and what makes it work is an efficient plan (Vaziri and Bakhshalizadeh, 2016).

Museum is a Greek word inspired by "Museum of Ion" meaning House of Angels. Museums now play a special role in the development of the tourism industry. In the meantime, the existence of historical museums that in addition to attractiveness can be an important factor in introducing the culture and civilization of the host community (Shoaei & Ahmadi, 2016).

2.2.1 Statistics and Surveys of Communities have Museums

Statistics show that in the nineteenth century only a few people visited them. At its ninth session in Delhi in 1956, the UNESCO General Conference emphasized that it would be easier to visit these places, especially for the working class and consider the ways in which museums can be enriched. Other statistics show that countries with a variety of museums that are free to visit are visited by only one in every 200 people who go to the cinema. In the UK, it is customary for museums to make up for this shortfall if the school or any other educational institution does not devote time to teaching art. Some experts recommend establishing children's museums that are more attractive.

The museum has many daily visitors of different ages with different cultures. If the dedicated museum design and appropriate guidance determine the historical and artistic value of each object to the viewer, a simple pottery object from the first millennium BC will be as attractive and popular as a contemporary painting.

Museums have no restrictions on the display of works of art, natural, cultural, historical and etc. They are of various types, including the following:

Historical museums

Scientific Museums

Specialized museums

Fine art museums

Stamp museums

Technology museums

Botanical Museums

Hybrid museums that sometimes maintain a combination of museums (UNESCO).

2.2.2 The most visited museums in the world

Some of the most famous museums in the world include the Louvre in Paris, the Hermitage Museum in St. Petersburg, and the British Museum in London. Enthusiats around the world visit more than 10 museums, including:

Louvre Museum in France

National Museum of China

American National Museum of Natural History

American National Aerospace Museum

Museum of the Great Britain

New York Metropolitan Museum of Art

Vatican Museums

Shanghai Museum of Science and Technology in China

National Gallery of London, United Kingdom

Taipei National Museum Palace, Taiwan

There are other famous museums such as the Russian Armitage in St. Petersburg, the National Museum of Reina

Sophia in Madrid and so on which are less in number of visitors than the above ones.

2.2.3 The most famous museums in Iran

Among the most important museums in Iran, we can mention the following: Museum of Ancient Iran, Museum of Azerbaijan, Reza Abbasi Museum, Museum of Glass, Carpet Museum, Cultural Institute of the Foundation Museums including (Time, Money, Ramsar Palace, Historical Cars of Iran), Museum of Contemporary Art (established by Empress Farah Pahlavi), Palace Niavaran Museum and Safir Office, Bahabad Museum of Natural Sciences of Yazd are also the most famous museums in the field of natural sciences.

2.2.3.1 Private museums of Iran

Besides the public museums that we see in Iran, a number of private museums are set up and run by different people as well. In Iran, the largest type of private museum that experts believe is the largest private one in the Middle East, was launched by Dr. Mohammad Sadegh Mahfouzi. In this museum, which has been established for more than twenty years, more than 120,000 historical works from prehistoric, Elamite, Achaemenid and other historical periods, as well as works from the post-Islamic period are kept.

The first museum in Iran

Throughout history, there have been places to keep some important and beautiful old artifacts. Examples from the Achaemenid period can be seen in documents that were private. According to current definitions, the museum was established about 120 years ago in Iran during the reign of Mohammad Ali Shah. He turned a part of Golestan Palace into a museum to display precious, royal and historical tools and exposed it to the aristocracy. There are many museums in Iran now, but their number is very small compared to the size and history of Iran.

2.2.4 Classification of Museums based on the form of presentation of works

Familiarity with various museums
Museums are classified in different ways: Museums of History and Archeology, Outdoor museums, Anthropological museums, Palace of Museums, Museums of Science and Natural History, Regional-local museums, Traveling Museums, Museum Park, Military Weapons Museums, Museums of Thinkers, House of Artists, …

Outdoor (Open Air) Museums
These museums, as they are known, are formed in the open air. By creating these types of museums, it is possible to introduce the important findings and knowledge of archeology. When a scientific archaeological excavation leads to the desired results and the discovery of valuable immovable artifacts and can not be transferred to museums, by providing the necessary conditions and facilities, it will be prepared as the desired place for public viewing. This is known as the Open Air Museum. Outdoor museums are divided into two categories: eco-museums and site museum. Eco Museum is a wallless museum that creates and develops a new cultural-historical tourism industry for an area. Eco Museum is a living natural site that displays the distinctive and valuable nature and identity of an area. Through it, local communities introduce, interpret, manage and protect their cultural or natural heritage (Akbar Beigi, 2013).

Museum Park
Due to having various scientific and cultural dimensions and recreational and educational attractions as well as recreation,

they are of great importance because they show closely the biological and natural image to the people. An important feature of these museums is that the general public can benefit from visiting them. There is no history of creating a museum park in Iran, but it is common in countries such as China and North Korea (Raessi, 2016).

Street Museums

Street museums can be seen in many squares and passages, replicas and large images of decorative objects and motifs, historical monuments, plastering and scenes of historical events that have been reconstructed in the intended delegation or embodied on bronze walls and plaques. The most interesting example of this is the construction of architectural spaces in the form of cramped pottery in a square in Bukhara uses empty space as a restaurant (Shamse Alam, 2015).

Virtual Museums

They are formed for the rapid advancement of cultural goals and due to the lack of facilities available in deprived areas and cities. These museums display different cultures in different places. If enough attention is paid to these types of museums, they will be very impressive (Heidari, 2013).

2.2.4.1 Museums Classification according to the Audience

Museums, according to their mission and goals, are to present objects and information obtained about them and to make connections with concepts and findings. In fact, they are a bridge between these concepts and the audience. Today, audiences and visitors to museums form different groups and categories of society who want to find answers to their questions from this cultural center. In general, we can mention primary school students, adolescents, puplis and age groups 18 to 30 years, adults, family groups, foreign tourists, specialist groups, physical and motion disabled, ethnic minorities, mixed age groups. Today, museums are shifting from the object- to the audience-oriented, which has doubled the knowledge of the museum audience. In this way, psychologically and consciously dealing with the audience and recognizing the behavioral spirits of the visiting groups will help the museum owners and guides to achieve their goals. Therefore, in the first stage, the museums implemented the introduced patterns by recognizing their audience. What knowledge the audience came to the museum with and what is the purpose of their visiting, is important for the museum guide to be able to convey his/her information and knowledge to the audience. At this point, the guides start asking questions of the audience, before the audience or the visiting group asks their questions (Akbar Beigi, 2013).

The level of awareness of people about the items displayed in the museum varies. For this reason, the knowledge of the audience is first extracted and then the museum guides are addressed to the audience by special methods to replace the information or reconstruct it. This action requires different time and energy for different people. It is very important to ask questions about museums and amphitheaters that motivate more people to visit the museum many times. This can be achieved by the unusual display and arrangement of objects and the use of questionable signs and symbols. There must be a cultural dependance and love between objects and visitors and the museum was used by everyone as an educational and research complex. Children's Museum, Museum of the Blind, Museum of the Disabled, Museum of Women, Museum of the Elderly, etc. are museums designed and built according to the needs of the audience (Lublich & Tapan, 2015).

Palace of Museums

This type of museums are historical buildings that have come to us from the past and represent the way of life of their owners (Jodat, 2002). A building may also contain historical objects and works of art, including murals, carvings, and plastering. Palace museums are usually built in governmental centers. The purpose of establishing these museums is to display historical works and monuments, as well as to learn lessons. Chehelston Museum Palace, Saad Abad Palace Complex in Tehran and Malek Abad Garden in Mashhad are such museums (Shaykh al-Islami, 1999).

(Figure 1). View of Saadabad Museum Palace. Study sample, author's source, 2020

Museums Divisions by Scale

In this category, museums can have a variety of themes that are divided into the following categories:

Museums, global or international, whose collections whatever they may be, represent that theme in all parts of the world like the Metropolitan Museum, which deals with the history of art in different parts of the world (Heidari, 2013).

(Figure 2). Iran National Museum, Source, Khosravi 2015

Regional teachings, like national ones but on a more limited scale, are usually subject to the political divisions of the country, which include the provinces and states of that country, and represent specific regions.

Local teachings that are on the smallest scale and represent the culture of a particular region or neighborhood and they only display historical artifacts and objects from the same area (Akbari, 2016).

2.2.5 New Theories of Museum Building

New theories of museum building, which led to changes in important museums such as the Louvre, were the result of a revision of popular attitudes toward this type of space. In practice, it is proved that the combination of specialized uses with ordinary attractive uses can play an important role in the success of the museum. And that new museums, whether scientific, artistic or cultural, are no longer places that only wise people can visit for specific purposes. Rather, they are spaces for spending leisure time, which at the same time make it possible for people to get acquainted with serious and main themes. In fact, the accepted tendency today is to combine cultural and scientific activities with recreation, entertainment and social enjoyment (Sylvester, 2015).

Scientific and cultural museums, painting galleries, exhibitions, rather than providing original examples and detailed experiments to the attention of experts, mainly dedicated to educational-recreational programs, game-like simulations, showing many replica, diverse and entertaining replicas and examples. Today, museums are no longer viewed as a sanctuary or a temple and a place of art. Today, museums have become very much like leisure centers that offer a variety of activities, from selling food to shopping and doing a variety of activities, something like a recreational park (Akbari, 2016).

Duties of Museums

A museum should not be considered a place in which only historical and archeological works are displayed, but all art, scientific, animal, medical, galleries, libraries and archives and most historical monuments are a kind of museum. Any object and work exhibited in a museum or exhibition has a language and communicates with its viewer. By meditating and thinking, the present language of these works can be understood and examined from different perspectives. One of its most important tasks is to establish a cultural connection between the visitor and the object on display. In fact, one should try to convey to the visitors the same connection and feeling that existed between the creator of a work and themselves, which is not far from reach.

Museum design with a traditional architectural approach

The phenomenon of traditional architecture as a category in aesthetics and mysticism is very important in the purity of thought and respect for nature. Traditional architecture, although it has undergone transformational phenomena throughout history, has been able to maintain its special identity and represents customs, mood, feelings, tastes and art. A museum is an institution that collects the study and preservation of objects representing nature and mankind in order to spread awareness, education and pleasure in front of everyone. By definition, the term museum not only includes the institutions bearing the name, but also includes art galleries, non-commercial galleries, art galleries, religious and non-religious treasures, some historical relics, and permanent exhibitions in open air and so on. This shows the importance of designing a museum in a traditional place (Daadras & Lazempour, 2017).

2.2.6 Museum Architecture

Like the museum, the architecture is largely from the past. The architecture of any museum is subjected to its history. Medieval religious treasures are kept in churches and monasteries and royal complexes, in palaces or castles, the role of art with the housing of a class of society that gave rise to it. The connection between the two is close. As the remnants of feudalism gradually evacuated during the nineteenth century in favor of the establishment of political and cultural democracy, the palaces of many institutions of the former system became museums. While in the lands or cities without new palace museums, more or less carefully, they were built imitating temples, palaces and castles. This is how the first major US museums turned to "neoclassical" or "Palladian" architects. Museums established by the British in India, became the works of "Neo-Gothic" or "Victorian" architecture. In the Soviet Union after the 1917 revolution, royal and aristocratic governments became houses of culture for the people.

The myriad monuments that were ransacked by that government or local authorities following their liberation by their traditional owners or their confiscation by the new political regime forced the authorities to find new uses for them and turning them into a museum became a common way to preserve and revive them. Since World War II, changes in museological theories have led to evolutions in museum architecture, at least in the developed world. During the national and international conference, experts tried to present a doctrine about efficient architecture for museums (Raessi, 2016).

The vital role of museums in human societies is an innovative, lasting and promoter of the purest cultural phenomena. Museums are one of the few centers to preserve the relics of the previous generation and are in fact the children of art and history. The museum is the only place where objects are kept in order to raise the level of general knowledge, education, cognition and understanding. It is through these museums that the cultural heritage of the past is presented to present and future viewers, thereby giving a picture of the cultural past in order to achieve a fruitful cultural future. The museum viewer is curious about the objects on display and thinks about each one and how it came to be. The objects kept in the museums have some kind of cultural background. As cultural heritage, they embody corners of the history of human civilization and thought. Each object has a special value in itself and each viewer looks at it from a specific angle according to their knowledge and expertise. They examine it in their own way and acquires new knowledge (Jafarpour et al. 2016).

In our modern society, the museum represents the institutionalization of public tendencies to collect. One of the reasons for building museum can be considered as trying to fill the spiritual gap created after the Renaissance. The rapid movement of human beings caused a sense of anxiety about breaking of the past. And on the other hand, a kind of archaeological thought, question and approach was formed in the West that the attempt to reach the essence of everything, in search of a new meaning of human history and existence. But given that the culture and history that came from the past, belonged to the past, they were given to the museum. And historical art and cultural heritage replaced history and culture as an absolute tool of cognition. Therefore, the preservation and protection of cultural heritage and respect for it was the responsibility of a non-profit social institution called a museum, which required the creation of a database with a suitable body and appropriate to the thematic requested content. It is clear that success in this matter in the general view involves addressing the two main categories in the form of problems. The first is the maintenance of objects and their requirements, and the second is to deal with the architecture and quality of space and the quality of display of objects and their requirements. For this reason, the factors of optimization and promotion of the second category have always been the concerns of designers and museum owners (Khosravi, 2015).

2.2.6.1 Siamese Historical Museum of Japan

Attention to nature and its elements such as water, light, rocks and the like that can be seen in ancient architecture as well as in some contemporary Japanese works, as well as the design of the water museum or other museums related to important elements of nature and the tendency to the simplicity of the elements in the design of the museum space is rooted in ancient Japanese culture. The Siamese Historical

Museum in Osaka, Ando, is a magnificent monument that demonstrates the power of water to engineers who want to tame it. This building is located next to an artificial lake that dates back to the 17th century. Over the centuries, monks and feudal lords used their skill to build lakes, canals, or wooden or stone pipes to carry water to adjacent areas. The effects of the first engineering and civil activities have been obtained following recent excavations, especially due to the expansion of the coast and activities to maintain water flow (www.arcspace.com).

(Figure 3). View of the Siamese Historical Museum of Japan

Part of the old dam with a height of 15.4 meters was carefully examined. For this purpose, it was dried to reveal how different layers are added to each other. To place these works, Ando has built a multi-level space next to the reservoir (basin).

(Figure 4). View of the Siamese Historical Museum of Japan

*(Figure 5). Plan of the Siamese Historical Museum of Japan
(connection ramp)*

After crossing and climbing a communication ramp, a large concrete arena without aperture appears. Stairs go down from one corner of the arena to the courtyard and the visitor, after passing through it, finds themselves in front of a waterfall that falls from the walls and is located in the direction of a large water intake. This space is one of the attractive parts of the museum.

In the middle of the hall, there are parts of ancient canals and pipes that may attract the attention of archaeologists. But architecture students pay more attention to the glorious architecture of the museum.

(Figure 6). View of the forum channel

(Figure 7). View of museum

A corridor behind a curtain of water is designed as a promenade for visitors, the space is very pleasant due to the sound of water falling and the passage of desirable light.
At the end of this corridor enters a cylindrical volume that the visitor enters the museum leads to the silence of the sound of water and a sign of entering the museum space.

(Figure 8). View of the plan (entrance to the museum space)

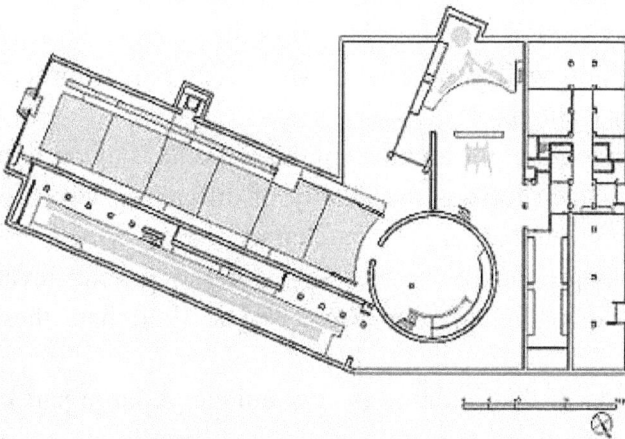

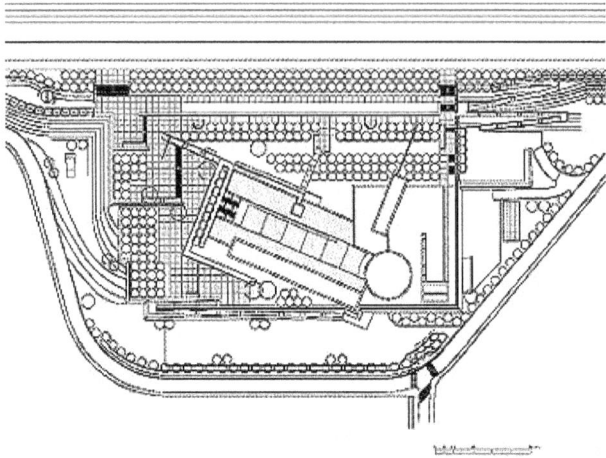

(Figure 9). Planning view of the Siamese Historical Museum of Japan

The simple volume is cube-shaped. It is formed with ramps and stairs and its materials include concrete, wood, iron, steel, glass, and so on. All this to play with water and light.

A long section of the work is taken through the main gallery space, is the deepest part of the project structure and is the key point of analysis in determining the penetration of natural light.

The short section on the right shows the changes in depth and height as well as the spaces where light passes through the water features.

The main focus of the architect of this building is to use light directly, which reflects the quality of different surfaces of raw concrete and areas of water characteristics.

Since a large part of the building is below ground level, skylights (sky lights) have been used to illuminate these parts.

The design of this building is very unique. A large part of the construction of the site is performed linearly by one of the visual features of water. This space has the potential to have very different quality at different times of the day. Due to not disturbing the landscape of the area, the wall is a short distance from the ground and is made of granite. After crossing a path along the water next to a wall of granite blocks, they reach the concrete square. From here the visitor at the bottom of the stairs, under a calm pool of water on the upper level that reflects the rectangular volume, reaches a water field with a waterfall on both sides.

(Figure 10). Three-dimensional view of the museum

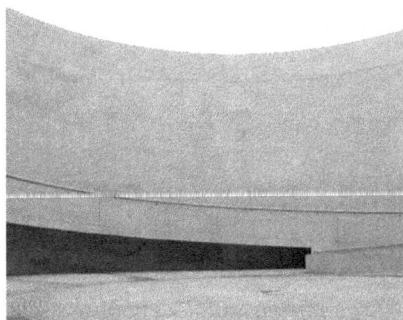

(Figure 11). View of the corridor of the museum

The main volume of this building is an exhibition hall, which is formed by the main dimensions of the museum's archeological excavations.

(Figure 12). View of the exhibition hall

2.2.6.2 Fort Worth Museum of Art by Tadao Ando

(Figure 13). 3D plan of the Fort Worth Museum of Art by Tadao Ando

This space has only one life when people enter. Therefore, architecture can play an important role in it. This space plays a role with architecture, which strengthens the exchange between people and ideas from paintings and sculptures, and more importantly, between the people themselves. The Fort Worth Museum of Modern Art is an example of the Japanese architect Tadao Ando with its simple geometry in a natural environment with minimal selection of materials. Five long flat roofs in single buildings of more than 1.5 hectares in side of which there is a swimming pool are other memorable projects of Ando.

This building is made of concrete, iron, aluminum, glass and granite. The museum is built entirely around a swimming pool. The beautiful trees and rocks around the museum are an example of Ando architecture. With its simple and unaffected design, this museum is considered as a modern work of art. Its surroundings are like a museum work of art, intertwined with the space of its large windows. Glass and water are in agreement with each other. When the calm pool shows the spaces, the glass also shows the water. Using glass as a wall, there is a physical barrier but there is no visual boundary between outside and inside. It is also light that causes water to be seen through the glass and its border disappears. The use of concrete refers to the feeling of Tadeo Ando on the surfaces and edges.

The edges are so clear that the material allows it. Large flat walls show us the basic structure of the building. Compared to the natural environment that surrounds its architecture, two elements make it more dynamic. The Fort Worth Museum of Art shows the highest and most elegant emphasis on its boundaries using materials to create architecture up to the surrounding pool. Light is also a key point in the design of the museum with an emphasis on scattered and reflected natural light.

Interconnected concrete roofs support sloping windows that provide natural light. The five Y-shaped columns, 40 feet high, support the sheets on it and are a sign of the museum. This museum is close to the Louis Kahn Campbell Museum of Art and the Amon Carter Museum, built by Philip Johnson. And includes more than 2,600 modern and contemporary works of international art in 53,000 square feet of gallery space (www.archdaily.com).

The museum consists of several rectangular cubes combined in an L-shaped plan. So, it includes a reflective pool. As the building splits, a set of rectangular cubic volumes is displayed. This is Louis Kahn's strategy for the Kimble Museum of Art. Ando tried to respond positively to the adjacent valuable work. On the other hand, the linear form has left the possibility of conceptual and physical expansion of the building. The walls of the site are responsible for delaying the audience's understanding of the building. Numerous walkways in and around the building invite visitors to view architecture from different angles with rich perspectives.

Ando expresses his belief in the guiding role of architecture in the way and lifestyle of people and expresses his interest in a kind of architecture that inspires people. Use their own resources to move into the future. He shows multiple linkage paths and spaces in their designs. From inside to outside the building, several layers of space can be identified. First, the display spaces and galleries form between the concrete walls. Then, a space is created between the galleries and a clear glass wall, sometimes with stairs or with a space with walls from the ceiling to the glass floor, used to see the distant view of the city beyond the pool. Finally, the area beyond the glass walls consoled with the ceiling includes a Y-shaped roof retaining element that blocks direct sunlight on the artwork and it mixes with water from below, which extends to the vicinity of the glass wall of the building. These are all arrangements by which Ando has been able to place the museum in a glass cover, unlike many other museums. The entrance is a simple opening and recess in the simple view of the side of street. Then there is the light-filled lobby, with transparent walls at the entrance and opposite. The galleries have various light sources and various heights, which sometimes cover the whole height of the building or half of it. The ceilings are also located on Y-shaped arches and linear slits along them allow natural light in. The light, controlled by flat and curved surfaces, creates arched interiors reminiscent of the Kimble Museum of Art. The reflection of the building in the dark at night in the reflecting pool is reminiscent of large luminous lights. Concrete, glass and water are still elements that Ando has shown in its beauty.

(Figure 14). View of the Fort Worth Art Museum from Tadao Ando

(Figure 15). View of different parts of the Fort Worth Museum of
Art by Tadao Ando

(Figure 16). Plan of the Fort Worth Museum of Art by Tadao
Ando

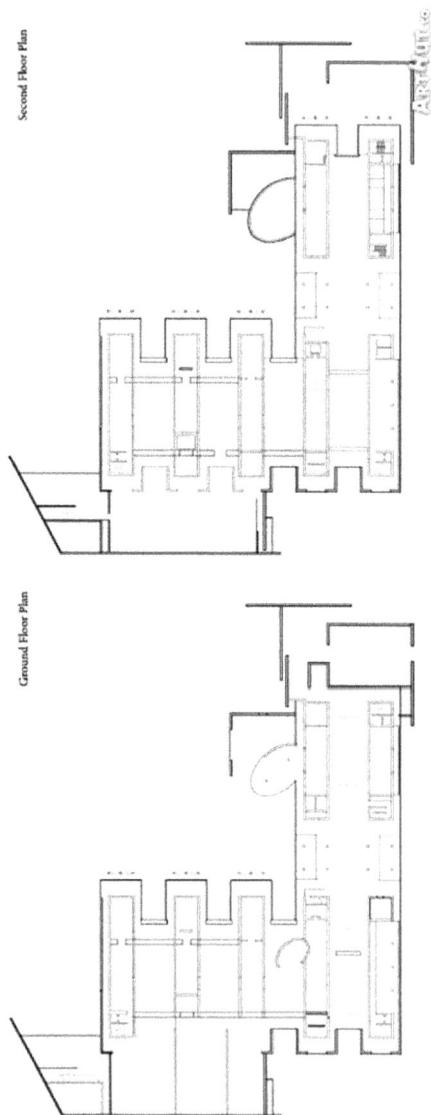

(Figure 17). Ground floor plan

Elevations & Sections

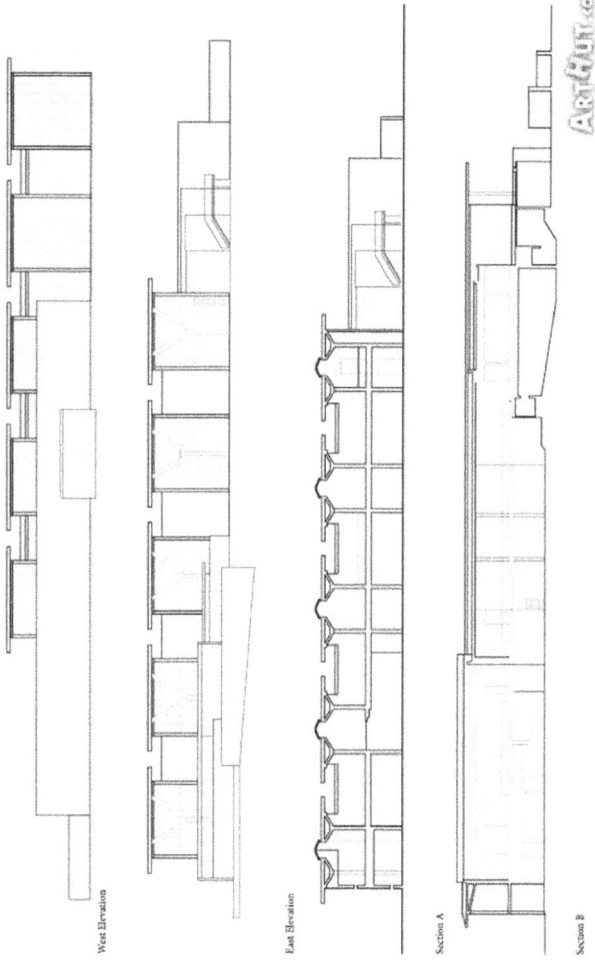

West Elevation

East Elevation

Section A

Section B

(Figure 18). Height and different parts of the plan

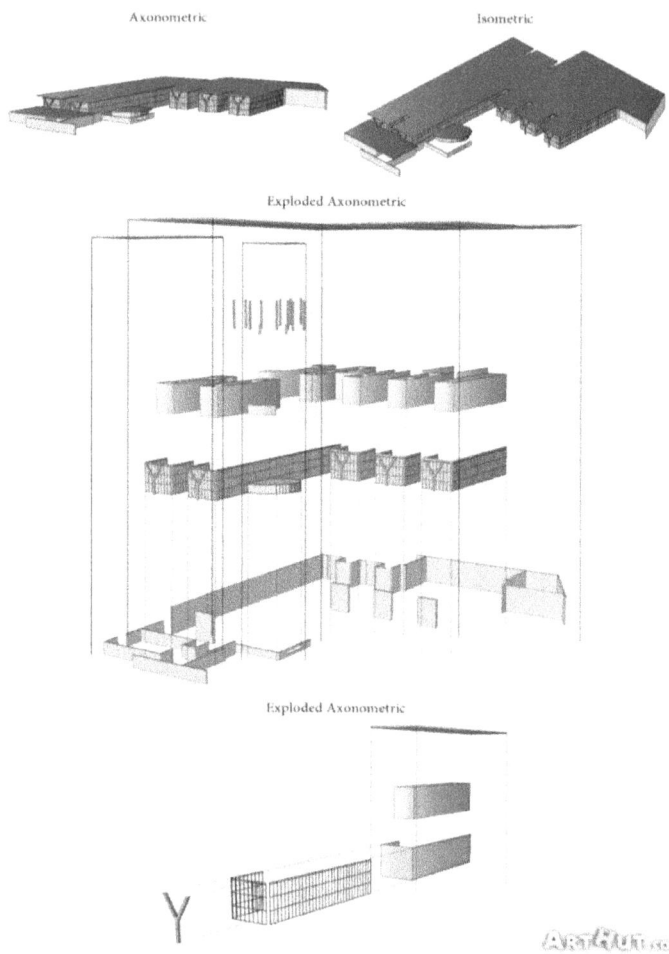

Axonometric

Isometric

Exploded Axonometric

Exploded Axonometric

Y

(Figure 19). Indicates the exonometry of the façade

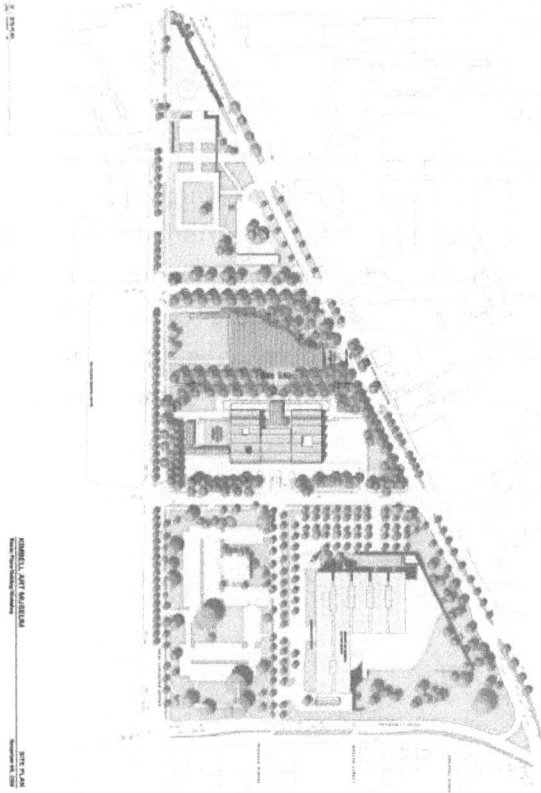

(Figure 20). Overview of the plan

This space has only one life when people enter it. Therefore, architecture can play an important role in. This space plays a role with architecture, which strengthens the exchange between people and ideas from paintings and sculptures, and more importantly, between the people themselves as well. The Fort Worth Museum of Modern Art is an example of the Japanese architect Tadao Ando with its simple geometry in a natural environment with minimal selection of materials (www.arthut.com).

2.2.6.3 Gholhak Water Museum Garden

The water museum garden is one of the contemporary examples of Iranian landscape architecture that has been built in Gholhak in recent years. The design of the water museum garden reflects the impact of water on human life and includes the garden and the aquarium building. The effects of water in this museum garden in the form of various fountains, taken from the architecture of other nations. Numerous spaces that have been created next to the fountains are connected to each other by angled axes to the park wall.

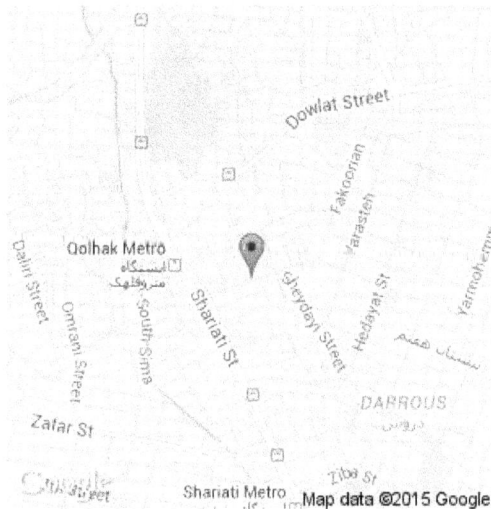

(Figure 21). Location of Gholhak Water Museum Garden

(Figure 22). Yakhchal St. not far from Ki Nejad St., (Sheidaei), Shariati St., Tehran

Water has long played a vital role in human life and lakes and rivers have always attracted people. In the recent past, when man used water as one of the components and elements of landscape architecture and urban design, its various effects have always stimulated and excited human toward beauty.

The question here is which aspects of the presence of water in human life and the environment and its role in wildlife and the natural environment of plants and animals can be shown? In an exhibition complex, various issues such as the evolution of irrigation and water supply methods, the role of water in transportation and communications, the role of water in agriculture, industry and mining, in production and energy, the role of water as a cleaner and detergent as well as the Earth's freshwater resources and ways to preserve and recover them can be offered and displayed. The placement of an old garden among the many buildings in part of Tehran, was an incentive for the municipality to organize and design it as a water museum garden, to preserve this old garden and to put it access as a dynamic part of the city to serve citizens and visitors. A garden that previously led to insecurity. The high price of land in this area as well as the lack of open land around it, especially lands with such a significant position, add to the importance of this piece of land. On the other hand, the existence of high environmental potentials (trees, aqueducts, surrounding gardens) and social (being among different educational centers) in the surrounding area made the important decision to use this piece of land.

The location of this old garden on the edge of the lively and active street "Yakhchal", an axis where different social groups can be seen, guarantees the use of the museum garden so that the more diverse the social groups, the more they use the set of growth. The garden of the Water Museum with an area of 8256 square meters is located in the northern part of Tehran and in a garden located on Yakhchal Street. After the correction, the area of the garden has reached 7161.9 square meters. This garden has a general slope of 2.85 percent from north to south, which initially had 382 trees and shrubs of sycamore, Tehran pine, mulberry, walnut, sparrow, etc., of which 190 has been used and the rest that were dry and sick have been removed. A stream of water, branching from a series of aqueducts, enters the garden from a northwest angle, and after a third of the length of the ground while rotating eastward, passing through the width of the garden. This project was put into operation in the summer of 2001 after 5 years.

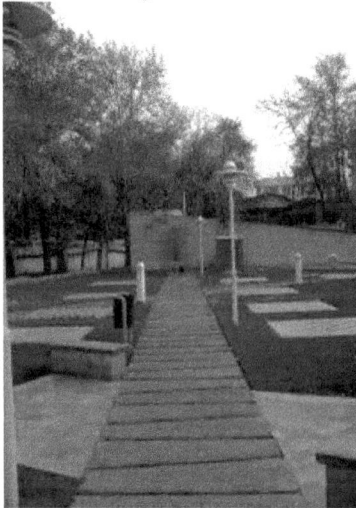

The other type of axes are recreational movement paths that have been constructed between the two types of mentioned paths. They are the only way to walk, stand, talk to friends, and sit down and so on. By equipping them with various urban furniture and attractive fountains, an attempt has been made to create a refreshing and attractive route. The shape of the path and its width and flooring are not subject to a specific law and vary depending on the location of the design. But in general, it can be said that in the design of flooring, an attempt has been made to draw the green space of the gardens in to the route and a combination of stone materials and grass can be seen. These paths sometimes end in a sitting area and sometimes stop in the surrounding gardens and do not connect to another path. This also strengthens the possibility of a person's connection with the natural environment. However, at the same time, there is a threat for people to pass through the grass, flowers, and similar spaces. The main building of the complex is the museum building, which according to the designer of the complex, has been very sensitive to its location. Because this integrated mass had to be built in a place that would cause the least damage to the valuable trees in the garden. On the other hand, it could not be formed at the entrance of the garden because not only did it block the view of the visitors to the whole garden complex and limit their view, but also the hiding of the building in the heart of the garden trees can lead people inside the Garden and rotate them in open spaces in front of the building. At the same time, as the building rotates relative to the main wall of the garden, a complete perspective is created from the point of entry and along the main axis of motion. While the end of this main axis ends with a natural element (falling waterfall) instead of an artificial element (museum building).

(Figure 23). View of the entrance to the Garden Museum

Water Museum and Exhibition is a three-storey building with a square cross section with dimensions of 21.60 meters. On the ground floor, which is 0.45 meters below the natural floor of the park, a glass and conical fountain is projected that directs part of the existing aqueduct water from the northern part of the first floor into it. Water is then transferred out from under the floor of this surface. In the northern part of the fountain, there is a multi-purpose hall. In the central core of the first floor, there will be a central gallery where the visual parts of natural effects and features are displayed.

The idea of designing the museum building was taken from the form of old conical glaciers with a wall in front of them to create shade. Its glass body evokes the symbol of ice pieces sliding on top of each other. This feeling is intensified by a rotation towards the concrete grid of the facade. The main entrance of the museum is stretched from Tria through a ramp located along the axis. Under this ramp, an amorphous fountain entrance is located so that the ramp acts as a connection bridge through which a person arrive from the garden area at the museum building.

(Figure 24). Three-dimensional plan of the garden museum

(Figure 25). Three-dimensional plan of the garden museum

Existence of various sitting places in different places of the garden, such as under the glass tunnel, behind the waterfall, next to the fountains and the movement axes, and most importantly the edges of the garden walls and along it, can create different and attractive views for the person. The equipped walls of the garden, inspired by the garden alleies of Shemiran, have created a suitable place for people to sit anywhere and with any desired view of the garden.

(Figure 26). View of the museum seats

(Figure 27). View of the stone platforms of the museum, which are embedded in a complex and group

In addition, there are complex and group (square) places with stone platforms where people can sit around each other and talk and relax. Creating cozy spaces to sit in the right places in the garden is also one of the important points. However, the sitting areas are arranged in such a way that whenever a person feels the need to relax and enjoy their surroundings, they can reach it in the shortest distance.

In the garden, there are many fountains with their own story. Huge globe floating in water, water tunnels and canals, waterfalls, numerous ponds in different dimensions and fountains and pool are each a collection of fountains in different gardens from the Persian courtyard pond to the Citroen Park fountain and the Japanese pond. Azure tiles are used in the triangular pool and the Persian pond is surrounded by a classic Roman portico (Ghazizadeh, Seyyedeh Neda, PhD student in Architecture, University of Tehran). It has used the entrance axis, which borrowed from Iranian gardening, and has also applied the intersection axes of modern European architecture along with the post-modern architecture of the body and the wall. The designer mentioned the past architecture only in the form of using Iranian materials and arches and did not pay attention to the design bed, which is an old garden. The regular geometric lines in the design of the park are reminiscent of the principles of modernism. However, paying attention to the elements of classical gardening, such as the semicircular porch of the classical Roman garden and the walls and facades of buildings, which are taken from Iranian architecture and taking a pattern from them, as well as a new look at the elements of the park (water and plants) embodies the principles of postmodernism.

(Figure 28). Gholhak Water Museum Garden Fountain

2.2.6.4 Introducing the Palace Museum Garden

This museum garden is located on a land with an area of 69,000 and an infrastructure of 20,000. Main space: Side space museum: Qasr prison, which during its life has marked a part of the country's political history and many Iranian celebrities have been imprisoned or executed in, it will eventually become a national garden that will leave a lasting impression on the future instead of violence. User space:

cultural, exhibitional, and recreational of Qajar Palace in 1213 AH, equal to 1176 AP in the second year of the reign of Fath Ali Shah outside Tehran at that time with a large mansion including inside and outside, garden and pergola, which is later known as Qajar Palace. This three-story building had a hall that could be seen from all four sides of the surrounding landscape and four towers were located on all four sides and pictures of Qajar princes were painted on the walls of the hall. This palace lost its original function from the end of Nasser al-Din Shah's period until it was destroyed in 1284 AP due to heavy rain and flood. After the first Pahlavi came to power, with the increase in the number of protesters and political prisoners, the area of the Qajar Grand Palace was selected to build the construction of the first prison in a centralized manner. By choosing this place, Qasr Prison was built on 11 Azar 1308 with 192 rooms for prisoners, of which 96 rooms were collective and the rest were individual, with a capacity of 800 prisoners. During the second Pahlavi era, several buildings were added to the complex, including the so-called political building. Many fighters of the Islamic Revolution and opponents of the Pahlavi regime were imprisoned in this prison. After handing over Qasr Prison to the Municipality of Tehran, many ideas were considered by the designers to turn this complex into a cultural center. Finally, according to the resolution of the Islamic Council of Tehran, the transformation of Qasr Prison into a cultural center focusing on the history of the Islamic Revolution was on the agenda of the Tehran Municipality. By conducting extensive historical studies in Tehran, the boundaries and dimensions of the Qajar Palace Garden were determined and applied in the design of the complex and the destroyed buildings were identified and introduced in the complex using various identification methods. The two most important buildings of the complex, namely the Markov Mansion and the building known as the Political Prison, have been restored in accordance with the technical principles and have been turned into the same prison museum during the first and second Pahlavi eras and thematic galleries. In conceptual design and identity of this collection, by using the art and creativity of Iranian artists and benefiting from modern and digital technologies, parts of contemporary history are presented in a purposeful, attractive and effective way. In addition to creating Iranian gardens in the complex, it has created a cheerful and fun atmosphere for citizens and visitors. The story of the Palace Museum Garden is the story of the 200-year history of Tehran and the struggles and heroism of political prisoners and the rich concepts of the Revolution, which has been gradually added to the collection and added to its culture and content. The story is narrated by the building and the events inside it, which tells the good and bad days of its life. It also tells the story of the struggles of men and women who gave their lives for freedom. In the meantime, it reviews its memories to recount the situation and atmosphere of the story. Other parts of this 7-hectare complex, which has 20,000 square meters of infrastructure in the final stages, include the Palace Shrine, Zurkhaneh, Palace Library and Documentation Center, exhibition spaces, cultural product sales centers, coffee shop and garden restaurant of the Palace Museum. Also simulating the 200-year history of the buildings, from the garden to the palace, from the palace to the prison and from the prison to the museum garden, the reconstruction of the prisons of the first and second Pahlavi period in the Markov building in the form of a museum, creating a real and influential atmosphere of the situation of political prisoners in Qasr Prison, conveying the desired concepts through volumetric works and conceptual arts are examples of concept designs in the

Qasr Museum Garden. This complex will be ready for operation and reception of Tehran citizens in the late summer of this year.

(Figure 29). View of the Qasr Museum Garden

(Figure 30). exterior view of Qasr Museum Garden

2.3 Thermal Design of Museum Complexes

2.3.1 Air Conditioning

In the definition of air conditioning, it can be said that performing operations on the air in order to make the air conditions of the desired place for living, working or certain industrial operations comfortable and hygienic and bring them to the desired level. Air conditioning causes the air condition to remain constant or change automatically according to a certain method. Air conditioning or pleasant ventilationis a branch of mechanical engineering. Its job is to provide conditions that are led to human well-being and are needed to maintain a particular product or process. To do this, a device of adequate capacity must be installed and monitored throughout the year. The capacity of the device is determined by the maximum instantaneous load and the type of control is also determined according to the conditions that must be provided during the application of peak and partial load.

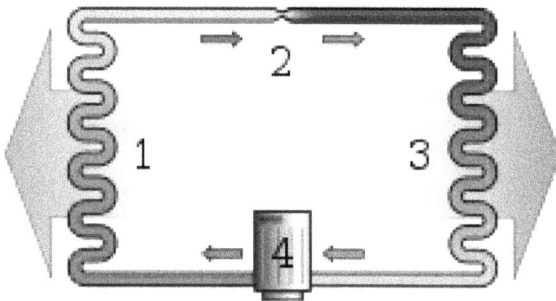

(Figure 31). A cooling cycle: 1) Heat transducer, 2) Thermal expansion valve, 3) Evaporator, 4) Compressor

Load estimation may sometimes be done accurately and sometimes by thumb methods. Accuracy in load estimation is one of the factors in optimizing energy consumption. Air conditioning usually includes cooling, heating, humidification, dehumidification and air purification (Ziegler et al. 2017).

In general, natural ventilation in a building has three different functions, which are:

Providing breathable air inside the building by replacing fresh outside air with polluted and consumed indoor air. This function is also called ventilation for health. Creating physical comfort by increasing the rate of reduction of excess body temperature by evaporation of sweat created on the skin as well as by removing the discomfort caused by wetting the surface of the body with sweat. This function is also examined under ventilation for comfort. Creating physical comfort inside the building by cooling the building material when the indoor air is warmer than the outside air. This function is considered as ventilation to cool the building.

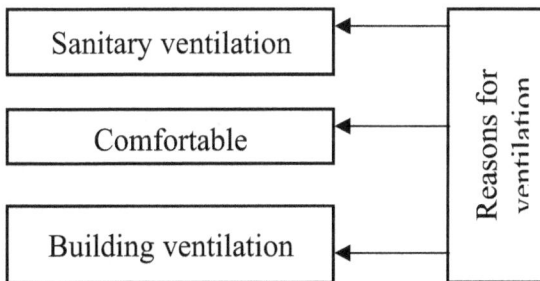

(Diagram 1). Natural ventilation performance in the building

2.3.2 Common Methods of Air Ventilation

Air conditioning/ventilation is usually done in the following two general ways:
Air conditioning: Artificial ventilation and natural ventilation, but more precisely, ventilation systems can be divided into the following categories:

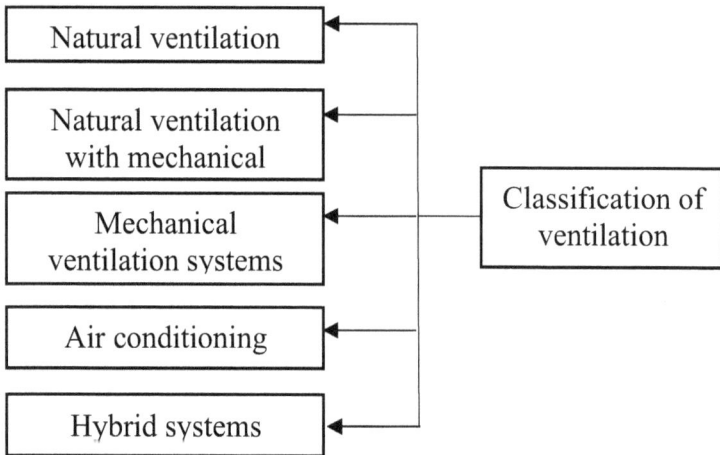

```
┌──────────────────────┐
│  Natural ventilation │◄──────────┐
└──────────────────────┘           │
┌──────────────────────┐           │
│  Natural ventilation │◄──────────┤
│  with mechanical     │           │    ┌─────────────────┐
└──────────────────────┘           │    │ Classification of│
┌──────────────────────┐           ├────┤ ventilation      │
│  Mechanical          │◄──────────┤    └─────────────────┘
│  ventilation systems │           │
└──────────────────────┘           │
┌──────────────────────┐           │
│  Air conditioning    │◄──────────┤
└──────────────────────┘           │
┌──────────────────────┐           │
│  Hybrid systems      │◄──────────┘
└──────────────────────┘
```

(Diagram 2). Classification of air conditioning systems in buildings

2.3.3 Natural Ventilation and its principles

Natural ventilation can provide fresh air, help transfer internal odors and heat, cool the structure and reduce structural radiation, as well as create evaporative cooling of the body and air. Natural ventilation occurs for the following reasons:
Wind air pressure difference;
Temperature air pressure difference.
The benefits of natural ventilation are compelling. Energy costs are significantly reduced, air quality is improved. Airborne chemicals are minimized by air conditioners or other mechanical devices. Overall, the use of natural

ventilation in the home can have a very positive impact on residents, the building itself and the environment.

2.3.4 Natural Ventilation Mechanism

Wind is an effective, determining factor of the movement of air inside the building. When the wind hits the building, the direct flow of air around and above it is broken and scattered. In this case, the air pressure is high on the windward surfaces (pressure zone) and very low on the backward winds (suction zone). Thus, pressure differences are created at different levels of the building (Figure 1).

When the wind blows vertically into a rectangular building, the front walls are subjected to pressure and the back walls are subjected to negative suction or pressure. If the wind blows obliquely into the building, the two opposite surfaces will be pressurized and the other two will be sucked.

The roofs of buildings are always located in the suction area. Of course, in the case of sloping roofs, this is true when the wind slope is low. Windward surfaces of steeply sloping roofs are in the pressure zone and their backward wind surfaces are in the suction zone (Figure 4).

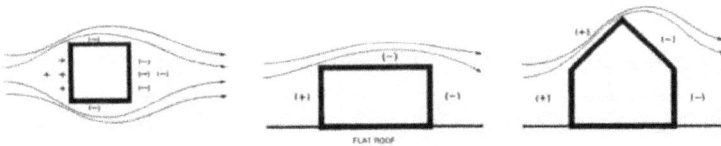

(Figure 32). Pressure and suction on the sloping roof. Figure No. 2: wind pressure and suction on flat roof Figure No. 1: the effect of wind on the building

The pressure difference that is created in this way on the surface of the building walls can be used to create natural ventilation and draught inside it.

2.3.5 Use of Natural Ventilation in the Past

The importance of wind in the design and construction of residential environments has long been considered. Aristotle four centuries BC and Vitruvius one century BC discussed the use of wind in architecture and urbanism. In our country, for many centuries, all buildings have been built according to the climate and environmental conditions. Sun, wind, humidity, cold, heat and in general climatic and geographical conditions have a direct impact on the traditional architecture of Iran in different regions. For example, to create a cool atmosphere inside hot desert houses, windcatchers, which is an innovative Iranian method, were used to make the living space bearable, which we will discuss how it works in the following.

2.3.6 New Methods of Using Natural Ventilation

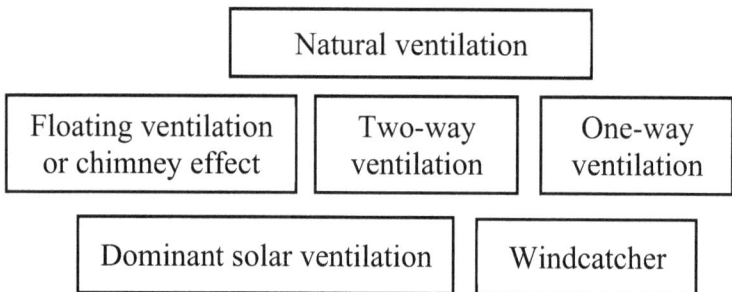

```
                    ┌──────────────────────┐
                    │  Natural ventilation │
                    └──────────────────────┘
┌──────────────────┐  ┌──────────────┐  ┌──────────────┐
│ Floating ventilation │  │  Two-way     │  │  One-way     │
│ or chimney effect │  │  ventilation │  │  ventilation │
└──────────────────┘  └──────────────┘  └──────────────┘
   ┌──────────────────────────┐  ┌──────────────┐
   │ Dominant solar ventilation │  │ Windcatcher  │
   └──────────────────────────┘  └──────────────┘
```

(Diagram 3). Classification of ways to use natural ventilation

2.3.6.1 Windcatcher

Windcatcher is an innovative Iranian method for creating a cool atmosphere inside hot desert houses. This air conditioner has made the living space of the Iranian people bearable for many years. Windcatchers are usually small turrets in the form of quadrilaterals or regular polygons, which are mostly higher than other parts of the house and are located on the roof. Windcatchers are generally built on a part of desert houses called the house-pond. The house-pond is a small porch that is located at the end of the summer rooms of each mansion. The windcatchers were located exactly on top of this pond and directed the air flow to the pond water through the vents (Figure 4). Existing studies in the field of natural ventilation of buildings show that the space of the inlet and outlet openings of the windcatcher constitutes 3% -5% of the floor space and when there is no air, the windshield acts as a suction device.

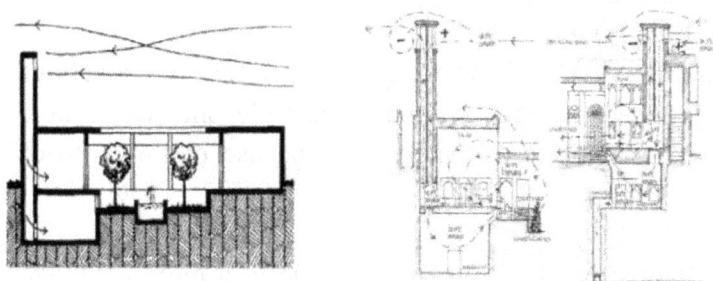

(Figure 34): Natural ventilation with windcatcher (Figure 33): Windcatcher operation mechanism

Windcatchers with different shapes have been designed and implemented in many central and southern cities of Iran according to the desired wind speed and direction.

2.3.6.1.1 Types of Windcatchers

The first type:
Windcatchers are of several categories in terms of external shape. The simplest type is a single-sided windcatcher. This type of windcatcher is made very small and humble above a chamber such as a heater hole in the roof. In this method, to avoid tornadoes and severe storms, the windcatcher is made only in the direction of cool winds and pleasant breezes, and the other sides are closed.

In some cases, one-way windcatchers are made back to the strong and annoying winds. In fact, this windcatcher performs the function of ventilation and evacuation of air. This example is mostly seen in Sistan and some parts of Bam cities.

The second type:

Two-sided type that has two sides facing each other and is made with long and narrow windows without protection and inside the building, it can be seen in the form of one or two holes in the niche. This example is seen in Sirjan and rarely in Kerman.

The third type:

It has three sides and has two types: three connected factions and three separate factions. In this example, one, two or three sides can be used separately, but the use of this type of windcatcher is rare.

The fourth type:

Four-way windcatchers that are made in a more complete and detailed way than other types. And usually inside its channels are divided into several parts with blades of brick or wood or plaster. This example can be seen in Yazd, Kerman and Bushehr, and so on. In Yazd and some parts of center of Iran, polyhedral windcatchers (usually octagonal and sometimes circular) are common.

The fifth type:

They form windcatchers like pipe (Chopoghi) windcatcher

The sixth type:

It is a windcatcher that the manufacturer has used to create several curved pipes for the external volume of the windshield instead of a cubic space of external shape. But internal channels and parts are like multifaceted examples. This type of windcatcher has only been seen in Sirjan.

2.3.6.2 Other Ways to use Natural Ventilation

1. Ground coverings. Cover the pavement and the ground around the building with grass or materials that absorb less heat. For example, by choosing green, the heat remaining in the ground is very small. This helps a lot in directing cool air into your house or building

2. Hills. The hills are wonderful examples of nature-inspired, whose idea of air circulation is taken from the hills created by insects. They are great for adding as part of your home, garden and green space to help with air circulation, especially during the summer months.

3. Water elements. Like many green spaces and Asian home designs, water is a very important factor. Fountains, ponds or swimming pools can act as transitional spaces in which the air cools naturally before entering the indoor space.

4. Furniture made of weed of basket weaving or bamboo. These materials are able to improve good air flow due to their hollow structure and lower heat retention. Weed of basket weaving and bamboo are not only ideal for outdoor use but they work great indoors.

2.3.6.2.1 Design to Enhance Natural Ventilation

Air pressure and velocity are among the most transient parameters of air movement. Pressure stimulates the flow of air. The pressure difference between two points, like a driving force, moves air particles from the source to the destination. This displacement will continue until the pressure difference reaches zero.

When an element is in the airflow path, pressure is created on one side and decreases on the other. After examining the different walls, we find that parabolic forms can create the maximum difference in air flow. Parabolic forms are geometrically capable of trapping air in one direction. This feature increases the wind pressure on the concave side and naturally decreases the pressure on the convex side. As mentioned earlier, creating a pressure difference provides the driving force needed for air to flow inside the building. In the simulation, the wind with a speed of 6 meters per second on the parabolic wall with a span of 10 meters has created a pressure difference of about 300 Ps. In parabolic three-dimensional form, airflow is trapped in both the XY and XZ axes and the pressure difference intensifies. In the simulation of the three-dimensional parabolic form with the same conditions, a pressure difference of about 1167 Pascals is created, which is a much stronger driving force than the previous conditions (Figure 2). Therefore, three-dimensional parabolic forms can be used to increase the pressure in the building. When the parabolic form is on the opposite side of the wind, it will reduce the pressure on its concave side. Using these principles, low and high pressure points can be created by designing the form and the driving force required for natural ventilation can also be formed.

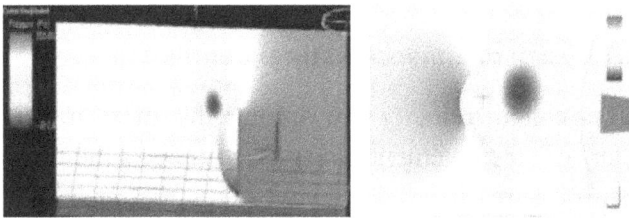

(Figure 35). Simulation in flow design software

The designed form should be tested by CFD calculations. In the analysis performed on FLOW DESIGN 2014 software, it is shown that there will be a pressure difference required to provide natural driving force for air to move into the residential space.

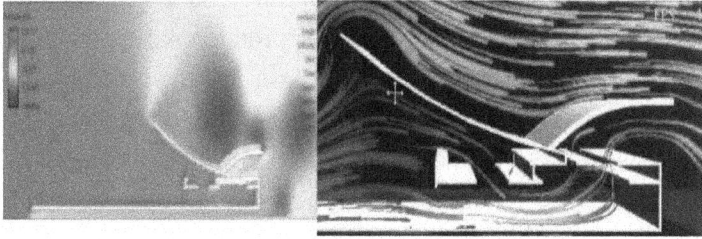

(Figure 36). Simulation in design flow software

In figure 2.9 the red color shows air pressure and the blue color shows the decrease in air pressure. As you can see, the pressure difference is well established and the image on the left shows how air flows inside the building. In the front part of the building, air enters the ducts of the building and enters the space of the house through the vents and through the back of the building, which has less curvature, it causes suction that air exits from the back of the building.

The air is always moving from high to low pressure. Creating high and low pressure points in a building block is made possible by designing the building form. U-shaped form of GSW building is the main idea of building design which is based on the principles of natural ventilation. We also designed a modern form of settlement based on the principles of creating high-pressure and low-pressure points and using parabolic forms that are suitable for the climatic conditions of the region. CFD analysis of the designed form shows that the function of the building will be suitable for strengthening natural flows and creating air draught.

2.3.6.2.2 Investigation of Physical Solutions of Natural Ventilation used in Past and Present Buildings

Natural ventilation is known as an easy way to exchange hot and unpleasant indoor air with fresh and air conditioned air outside the building. Since one of the biggest problems in the world, especially in developing countries, is energy and the lack of non-renewable resources, it is time to take a step in this direction. In this regard, the purpose of this study is to investigate natural ventilation solutions in the past and present, which has always met the needs of residents. This research has used the library resources, documents to address the issue under discussion. Finally, by gathering information in a table, ventilation solutions are classified according to time period (Rahai and Ghasemi, 2018). Natural ventilation is known as one of the least expensive methods of passive cooling, which reduces energy consumption inside the building without compromising thermal comfort and indoor air quality. Because green buildings have attracted special attention today due to their ability to provide a good level of thermal comfort and indoor air quality. However, these buildings consume the least energy compared to buildings equipped with mechanical ventilation systems. Natural ventilation can be created around or inside a building through wind or the buoyancy or combination of the two, known as natural convection.

Natural convection flow is one of the most effective ways of ventilation inside the building and occurs when wind and floating flow are in the same direction. In order to facilitate the prediction of natural convection constant flow characteristics, Atrij classifies two of the main types of openings according to the L/d aspect ratio, where L is the length and d is the width of the room. The air moves due to the pressure difference and the wind causes the air to flow from one point to another and plays an important role in the thermal comfort of the house. Heat flow occurs when there is a difference between the density of the indoor and outdoor environment, which is also achieved by the difference between the temperature inside and outside the room. In the natural ventilation method, the action of moving the air is done through the effect of the chimney based on the movement of hot air upwards and the entry of cold air from below instead or through the air draught, which move the air through positive and negative wind pressure. The shape and location of the building (for example, being in an open or dense environment or being high or low) determines how the building is naturally ventilated. In general, ventilation can have a dramatic effect on indoor air quality. Hence, ventilation is an integral part of today's buildings. Sandel states that humans are mainly exposed to indoor air whereas this is important in comparison with issues such as energy consumption, sustainability and air quality outside and pollutants.

In modern times, climate design is a logical response to crises caused by energy shortages and increasing environmental pollution due to fossil fuel consumption. Attention to climate is especially important for climates with acute living conditions. In general, three modes of ventilation can be considered: one-way, two-way, and chimney ventilation. Each of these modes shows the indoor air that needs ventilation and how it relates to the outside airflow. Natural ventilation is based on three climatic phenomena: wind speed, wind direction and temperature difference.

Wind speed: fhe wind flow in contact with the building creates a pressure field around the building. The amount of this pressure is based on wind speed. In this regard, ventilation is effective when the wind speed is more than 2.5 meters per second (9 Km/h).

Wind direction: The most basic factor determining how air flows through a building is the direction of the wind. When the wind moves over the building, it creates a positive or negative variable pressure field. Then the air flows from the positive pressure to the negative pressure areas.

Temperature differences: as the temperature increases, the density of the air decreases and the air moves upwards and is replaced by cool air. This phenomenon is known as the "chimney effect". The temperature difference between inside and outside the building and between its different areas causes a pressure difference and then air movement. Awareness of indoor airflows in closed environments is significant for three reasons: thermal comfort, indoor air quality and energy consumption of the building. Thermal comfort conditions are the range of temperature and humidity in which the body's temperature regulation mechanism is at its minimum activity. Factors such as air temperature, relative humidity, air flow rate, average radiant temperature, metabolic rate and body coverage are effective in determining the range of thermal comfort. Since natural ventilation regulates humidity and temperature, it can be concluded that ventilation is one of the most effective factors in creating thermal comfort. According to Olegi thermal comfort standard, the comfort range is between 21 to 27.7 degrees in terms of temperature and between 30 and 65 percent in terms of humidity. According to the US standard, this range is between 22.2 and 25.6 °C in terms of temperature and between 20 and 80% in terms of humidity.

Studies show that in the last two decades, attention to "indoor airflow" in the form of new knowledge has increased dramatically. For ventilation inside the building, especially in hot climates such as the southern regions of the country, it is better to have two-way ventilation and the air flow enters from one side and exits from the other side or the ceiling of the room. Without an outlet, even in the event of severe winds, there will be no effective airflow inside the building. Having an opening to the wind without opening the air outlet inside the building produces a pressure equal to the building pressure, which may even worsen the internal conditions and cause discomfort. In order to use natural ventilation, different solutions are used. Fans, wind chimneys, solar chimneys, single-sided or double-sided ventilated windows, double-skinned facades, and the use of a chimney-based atrium are some of the common solutions for natural ventilation. In natural ventilation, the roof is very important because the warm air tends to move upwards due to its lightness. In addition, the air flows passing through the roof are stronger and more stable. Also, natural roof ventilation is suitable for low-rise and enclosed buildings with low air flow around them, as well as buildings around which there is a lot of noise and pollution.

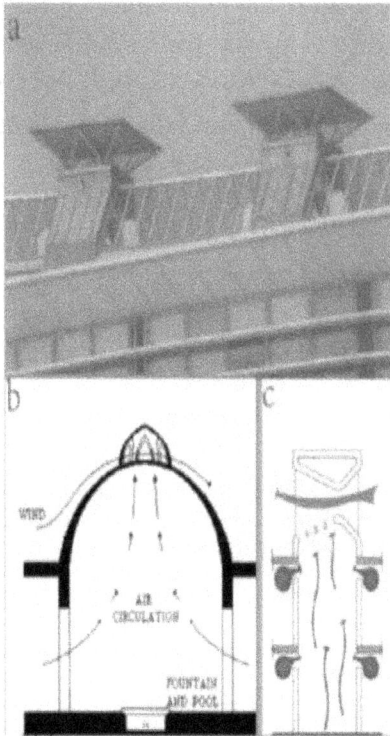

(Figure 37). Natural ventilation of the windmill

2.3.6.2.3 Windage

Windages are architectural components that are used to capture wind and draw fresh air into the building. The basis of the work of windage is based on air draught and they work like windmills on the roof of the building or as an independent part of the building.

These tall towers were first used in the desert areas of Iran and their function is based on bringing outside air flow inside, in order to reduce the temperature and establish thermal comfort conditions. In addition, positive wind pressure in the direction of the wind and negative pressure in the opposite direction of the wind provide these conditions. For this purpose, the inlet openings of windage must be toward the dominated wind direction. When there is no wind, these towers have the function of chimney ventilation.

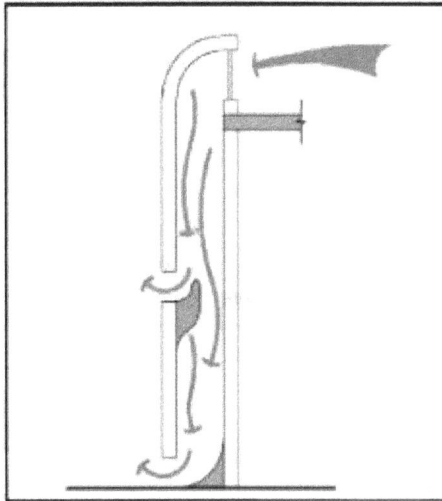

(Figure 38). Windages

2.3.6.2.4 Chimneys

Another common form of ceiling ventilation is chimneys, which are often cylindrical. Chimneys are based on the release of hot air from above and the replacement of cold air from below instead of. The operation of chimneys is independent of wind flow, but their outlet should not be where the air pressure is positive.

Numerous tests have shown the high efficiency of this ventilation system, especially in multi-storey buildings.

One of the active and sustainable methods in natural ventilation can be the direction and speed of wind, which should be used in buildings in such a way as to create the greatest potential in natural ventilation. The movement of air inside the building is also the result of the difference in air pressure between the two sides of the wind and the opening. When the air hits an obstacle such as a building, it puts pressure on the surface of the obstacle and creates a tornado. In such tornadoes, there is increasing pressure on the wind side "front the building" and at the end of the wind "behind the building". The placement of one of the openings of the building facing the high pressure area and the other facing the low pressure area allows the air to flow inside. Among the factors that affect these high and low pressure points in buildings are the principles of architectural design in environmental sustainability that are considered based on three principles.

2.3.6.2.5 Deployment and Arrangement of Construction and Urban Elements

In planning the shape of the city, arrangement and spacing can be used to achieve the desired movement. The establishment of construction groups and their effects on air movement is one of the most important issues. New buildings can be installed in such a way that the pattern of air movement is adjusted or modified to improve the ventilation conditions of other buildings. When the buildings are arranged in a grid, spacing usually ensures six times the height of the building air movement between them. The relationship between the height of the buildings and the space between them ensures the movement of air between them. The relationships between the height of buildings and the spacing between them affect air movement and have a direct ratio. The deeper and wider the building, the smaller the wind-protected space behind it. Forms of buildings create areas that are protected from the wind or divert it. This phenomenon will affect the patterns of air movement around the building and the comfort and outdoor conditions.

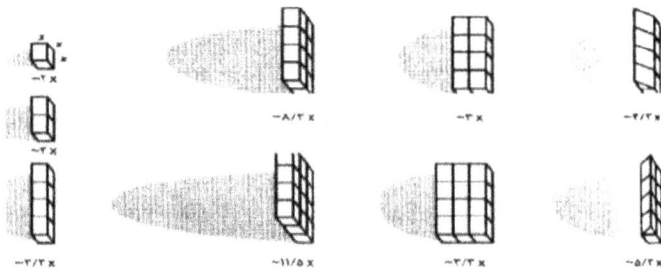

(Figure 39). The size of the space protected from the wind of the building

2.3.6.2.6 Peripheral Characteristics of Artificial Environments

Active design not only extends to the building but also extends to the surrounding space, where landscape architecture affects the climate inside the building. Architecture can play a vital role in changing the direction of air movement and should be directed in such a way that it passes through shaded areas instead of passing through the heated surface. In hot and dry climates, plants can be used to stabilize the movement of dust and reduce dust storms. In this way, the air is both cooled and refined before entering the building. A windcatcher can increase the difference in air pressure around a building. Exterior structures, such as adjacent walls, strongly affect the pressure created on the facades of the building.

These walls can be constructed using the dense greenery of a masonry structure. They can be used to guide the unpleasant wind and change its speed. For example, the side walls used at the end of the wind "behind the building" significantly increase the positive pressure created in the facade, if this event is in the direction of the wind and on the openings, then the opposite is true and we will face with pressure reduction.

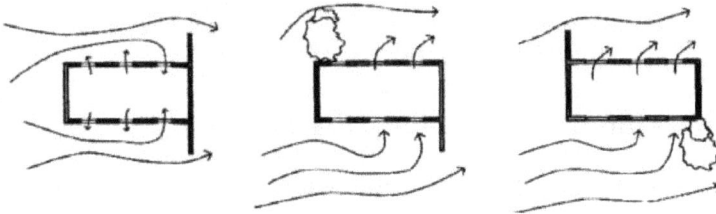

(Figure 40). Air guidance through the wall

2.3.6.2.7 Building Coverage

The principles of natural ventilation can be used in the design of the ledge, the shape and form of the roof, as well as its connetion with adjacent buildings. Therefore, extensive studies have been conducted on the impact of events that the potential for ventilation can be significantly increased. So, external protrusions that may also act as canopies can be placed vertically and horizontally to improve air movement. Outdoor air movement in hot and dry climates can cause discomfortaion due to comfort during midday hours because of warm weather. In such cases, it is necessary to reduce ventilation and prevent air moving of outside, unless the air is cool before arrival. Night ventilation is needed to cool the inside of the building and its occupants. For example: in shaping the building and covering it around the central courtyard of the herbal medicine plant, one of the ways of air conditioning is with the help of low openings from the central courtyard to the inside of the building.

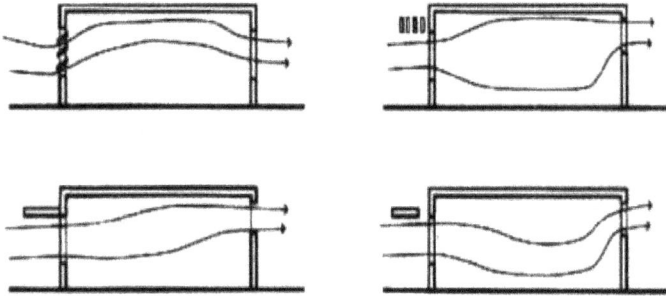

(Figure 41). Horizontal air flow

2.3.6.2.8 Interior Decoration of the Building

Improper design of the interior walls of the building can impede the flow of indoor air. When a wall separates a room or if several rooms are joined together, doors or corridors separate the air intakes and openings, then the air naturally changes speed and direction and comfort ventilation in the building is possible. But as long as the air goes from one room to another, the connection between the spaces is good and it remains open when ventilation is needed. However, the quality of ventilation in work or residential spaces does not only depend on the average speed or frequency of air change, and the fluctuation of these components in the space should be evaluated and it should be avoided the stagnant air cavities with no movement or change as much as possible. When the partitions are close to the entrance opening, the air velocity is at its lowest because the air has to change direction quickly. Each time the airflow hits an obstacle, its speed decreases. Where walls or partitions are unavoidable, it can be sured to some extent that of airflow as long as the walls start above the floor and end below the ceiling.

If the building has two openings in two different levels and areas (for example, an internal opening and an external opening) and there is also a temperature difference between these two areas, the effect of the chimney will be activated. Hot air naturally tends to rise and escape through the highest openings. The warm air that rises is replaced by the cold air at the bottom. This simple effect of falling cold and dense air and rising warm and less dense air is called the buoyancy effect and like the cooling agent, will help the chimney effect. The size, shape, and space between openings can be used to control ventilation, speed, and air volume. Then, when there is a difference in wind pressure outside, elements such as windmill, windcatcher or additional fan can be used to increase air movement and direct its flow.

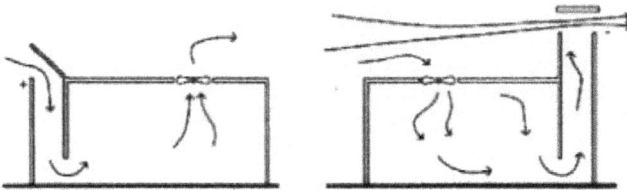

(Figure 42). The effect of air drop

Cooling in the building can be gained from natural resources in various ways. Here are two main methods used in hot and dry climates that we will describe:

a- evaporative cooling

b- ground cooling

evaporative cooling: air passing over water leads to evaporation, and as a result of such a process, heat is absorbed and air cools. Evaporated water is retained in the sky. Therefore, its humidity increases and evaporative cooling is suitable for dry climate. Directing the breeze from the pool or fountain before entering the building is the simplest principle. To ensure cool and humid air enters the building, the pool should be located between walls such as the central courtyard or, on a larger scale, in the middle of the building blocks, and the plants should be green to act as air purifiers and oxygen suppliers.

Ground cooling: we know that the soil temperature below the ground is equal to the average annual temperature. So, it is significantly lower than the air temperature during the warm season. The building cover is incomplete or complete in contact with the soil.

"Basements" produce cold due to the conduction between the soil and the wall. Finally, cold is transferred very quickly to the interior space and heat is absorbed by convection. Ground cooling gives the building the best use of heat mass.

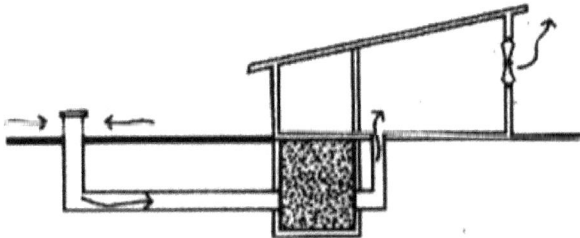

(Figure 43). Passive ground cooling

2.3.6.2.9 Natural Ventilation in the Past

windcatch is an innovative Iranian method to create a cool atmosphere inside warm desert houses. This air conditioner has made the living space of the Iranian people bearable for many years. Windcatchers are usually small turrets in the form of quadrilaterals or regular polygons in which the triangular structure is not seen at all. The windcatcher consists of a tower almost higher than the rest of the house, on the roof. Windcatchers are usually built on a part of a desert house called the house-pond. The house-pond was a small porch at the end of the summer rooms of each mansion. Summer rooms consist of large rooms with many doors. The house-pond is in the form of a space between the yard and the summer rooms. There was a small pond in the middle of this space and the reason for naming this space was due to the existence of this pond in the middle of this space. The windcatchers are located at the top of this pond, but they direct the air flow to the pond water through their holes. Windcatchers are generally made of clay and mud, and wooden beams have been used in the building to secure them against the wind. The windcatcher is decorated with patterned bricks. The windcatchers had inlet holes in the form of beautiful arches. How windcatchers work is in the form of water coolers today. In such a way that the wind entered it through the windcatcher holes and was directed toward the water basin as an assembly. After colliding with the pond water, evaporation was performed. Evaporation is a heat-absorbing action that cools the air entering from windcatcher opening. Then, the cold wind enters the summer rooms and cools the air inside. In some of the older mansions, which belonged to the wealthy people, the house-pond was a closed space, and the summer rooms had vents and corridors, such as air-conditioning ducts. Cool wind entered the rooms of the house through these corridors. Other uses of windcatchers were to cool the cellar space for food storage and to cool reservoirs. Windcatcher in Iran and other countries in the Persian Gulf operate both as wind turbines and as outlets. If hot air is drawn in due to a pressure difference from the back, the windcatcher acts as a chimney. At low wind speeds, the building continues to be ventilated only due to the chimney effect. In this case, the windcatcher can be considered as a type of solar chimney.

In the construction of windcatchers, the capabilities of Iranian architecture have been well utilized so that they, in addition to meeting environmental goals, are also part of beautiful impressive buildings. Passers-by crossing the city streets, watching various windcatchers, also enjoy the pleasure of sight. There are different types of windcatchers, but one of them is unique called the pipe windcatcher, which is currently an example of this type is remained in Sirjan. Some of the popular windcatchers are: windcatcher of Boroujerdi mansion in Kashan, windcatcher of Abbasian mansion (Abbasids) in Kashan, it was lower than the courtyard, windcatcher of Tabatabai mansion in Kashan, windcatcher of Dolatabad mansion in Yazd which is the tallest one with a height of 18 meters. 6 reservoir windcatchers in Yazd, as their name suggests are six. Badgir mansion, the jewel of Golestan Palace complex, a building belonging to the reign of Fath Ali Shah, which was built on the south side of Golestan Garden and became modern afterawhile during the reign of Nasser al-Din Shah with major conquests.

The name of this building was due to the existence of wind towers to produce cool, air conditioning and transfer it to the mansion, the house-pond and the main hall. Below the hall and mansion, there is a large house-pond with four tall windcatchers covered with mosaic, blue, yellow and black tiles with golden domes in the four corners to cool the air of the house-pond, hall and rooms.

The windcatchers had inlet holes in the form of beautiful arches. How windcatchers work in the form of water coolers today. In such a way that the wind entered it through the windcatcher openings and was directed to the water pool as an assembly and after colliding with the pool water, evaporation was performed. Evaporation is a heat-absorbing action that cools the air entering from the windcatcher opening. Then the cold wind enters the summer rooms and cools the air inside. In some of the older mansions, which belonged to the wealthy people, the house-pond was an enclosed space, and the summer rooms had vents and corridors such as air conditioning ducts. Cool wind entered the rooms of the house through these corridors. Other uses of windcatchers were to cool the cellar space for storing food and to cool water reservoirs.

Architecture has long sought the best and most efficient answer to climate needs. In general, the effect of architecture on the climatic performance of the building can be easily understood. Throughout history, architecture has always taken effective steps to provide or improve natural ventilation in line with climatic needs, following which thermal comfort is achieved for the residents of the building. Not only is it the most decisive answer to the needs of the past, but it can also be used in the modern architecture and serve as a light for the future. Therefore, the influence of the architecture of the building, the arrangement of the interior walls, how the spaces are separated, the height of the ceiling, the way the openings and the placement of the windows and even the color of the interior walls are the influential architectural factors in natural ventilation.

It can be seen that the need for ventilation and air flow inside the building has always been considered.

Looking at the studies as well as the results of this table, it indicates that some physical solutions of natural ventilation such as windage, building cover and interior decoration of the building, regardless of time constraints, have always helped buildings in natural ventilation as well as other physical factors such as the peripheral characteristics of the building, the placement and arrangement of building elements and chimneys are among the modern solutions used in buildings. Architects and manufacturers of buildings have always sought the most efficient answer to climatic conditions so that they can bring thermal comfort to residents. Because the principles that the ancients used for natural ventilation were efficient and effective; nowadays, they can still be used in modern buildings by modeling these methods as well as modifying their defects. Because of the urgent need for energy, resource scarcity, fossil fuels and global warming, it is recommended to use renewable energy such as wind to create ventilation and airflow. In architecture, natural ventilation is often recognized as one of the most energy-efficient environmental solutions for improving indoor microclimate and thermal comfort.

It can be admitted that the world today is facing two major misfortunes, known as energy shortages and global warming, caused by the overuse of fossil fuels. Hence, the focus on sustainable and efficient design is getting more attention than ever before.

Natural ventilation is known as a useful and important method to reduce energy consumption, which in turn leads to a reduction in greenhouse gas emissions, as well as creating an acceptable level of indoor air quality and maintaining health and comfort indoors. It should be noted that natural ventilation is considered by designers more than usual ventilation methods such as mechanical ventilation. In addition to reducing energy consumption and building costs, the designers of this type of ventilation system (passive system) have taken a step towards keeping the natural environment clean by reducing heat production from electric cooling devices and increasing the comfort of people through proper and permanent ventilation.

2.3.7 Basics of Ventilation

Ventilation is entering outside air into the interior space intentionally which is divided into two parts: natural ventilation and mechanical ventilation. Mechanical ventilation, which involves blowing air in by the fan and facilities devices, while natural ventilation is done without the use of facilities devices or using fossil energy for changing the air (Akbari, 2016).

2.3.7.1 Ventilation Purposes

Air conditioning of buildings pursues two main goals. The first purpose of ventilation is to provide the required indoor air quality (IAQ), which is basically place on fresh air and the removal or elimination of indoor air pollution. Another goal is the proper heat transfer mechanism. In fact, we do not have ventilation to store oxygen. This is because low oxygen is hardly possible. On the other side, concentration of carbon dioxide, is an indicator of man-made pollutants such as perfume and moisture, which is known as stale air.

Indoor Air Quality (IAQ)

Residents of a space have two basic needs regarding the air quality of their living room. First of all, the health risk of breathing air should be negligible, and on the other hand, the air should be fresh and pleasant and not old and annoying. The main pollutions are:

- Odor and moisture created by people and their activities, which often include CO
- Emission of building equipment
- Thermal comfort

Besides creating the right IAQ, ventilation plays a key role in maintaining proper heat.

Ventilation has three purposes during the day until it reaches the right temperature:

- Indoor air cooling by moving or removing heat from outside air until the outside air is cooler than inside
- Cooling of the building structure

- Attempts to cool directly the human body during convection and evaporation (Akbari, 2016).

2.3.7.2 Natural Ventilation

It is based on the flow of wind and heat as driving forces and is not a new phenomenon or invention. The use of natural energy to ventilation for thousands of years for humans and animals has been available to create suitable living conditions. The use of mechanical drive such as fans to provide natural ventilation along the ventilation ducts was realized during the twentieth century. The mechanical air conditioning system provided a constant airflow. This also provided facilities for air purification (such as air conditioning) and heat recovery. But despite the advantages of the mechanical ventilation system, the natural ventilation system has grown very well, which can even be called a period of modernity in the late 1990s (Cleon, 2010).

2.3.8 CFD

Natural ventilation is very important for energy conservation, reducing carbon emissions and improving the level of comfort around the building as well as indoor air quality. By simulating the wind flow around the building with a Computational Fluid Dynamics (CFD) approach, architects are able to combine architectural science and results of simulation to understand the project in more depth and analyze the strengths and weaknesses of the design. As a result, they plan the project with extra care or revise the current plans. In this paper, a study on the optimization of natural ventilation of the building was conducted by simulating the wind flow around the green building by CFD method from three aspects of site design, building shape and the outer shell of the building. More scientific ideas and suggestions are provided compared to empirical evidence that is consistent with architectural design and simulation analysis.

Numerical simulation of Computational Fluid Dynamics (CFD) of the current situation

The simulation was performed on a day of autumn and after using Gambit software to 3D simulation, Fulent software was entered to analyze the prevailing energy and viscosity. The gravity of the earth is 9.8 in the direction of gravity. In other condition, the entrance door and windows are closed and the fans in the window are assumed to be the outlet and the air conditioners to be the input. The equations used in the simulation are as follows (3-1):

(3-1)
$$P_{total} = \frac{1}{2}\rho v^2 + P_{static}$$

This is a turbulence model based on restructuring group theory (Wilcox, 1998). These results are in different constants of the samples in the standard k-ε model. In addition, there are other conditions and functions in the transfer equations for the kinetic energy of perturbation (3-2) and the rate of loss of that relation (3-3).

(3-2)
$$\frac{\partial}{\partial t}(\rho k) + \frac{\partial}{\partial x_t}(\rho k u_i) = \frac{\partial}{\partial x_j}\left(a_k \mu_{eff}\frac{\partial k}{\partial x_j}\right)G_k + G_b - \rho\varepsilon - Y_M + S_k$$

(3-3)
$$\frac{\partial}{\partial t}(\rho\varepsilon) + \frac{\partial}{\partial x_i}(\rho\varepsilon u_i) = \frac{\partial}{\partial x_j}\left(a_\varepsilon u_{eff}\frac{\partial\varepsilon}{\partial x_j}\right) + C_{1\varepsilon}\frac{\varepsilon}{k}(G_k + C_{3\varepsilon}G_b) - C_{2\varepsilon\rho}\frac{\varepsilon^2}{k} - R_\varepsilon + S_\varepsilon$$

The turbulent viscosity in this case is calculated using a differential equation in Equation (3-4):

$$d\left(\frac{\rho^2 k}{\sqrt{\varepsilon\mu}}\right) = .72\frac{\hat{v}}{\sqrt{\hat{v}^3 - 1 + C_v}}d\hat{v}$$

(4-3)

Here:

$$(5\text{-}3)$$

$$\hat{v} = \frac{\mu_{eff}}{\mu}$$

2.3.9 Indoor Air Quality (IAQ)

IAQ relates to the air quality inside and around buildings and structures as well as the health and comfort of the occupants of the building. It is also related to gases (including carbon monoxide, radon and volatile organic compounds), (microbial contaminants), molds, bacteria or any stressful mass or energy and can affect adverse health conditions. Source control and purification use of ventilation to dilute pollutants is the main way to improve indoor air quality in most buildings. In a residential unit, we can improve the air quality inside the building with very simple solutions, such as regular carpet cleaning.

The US Environmental Protection Agency has guidelines for cleaning based on traffic, number of family members, pets, children and smokers. Carpets and rugs act like an air filter and need to be cleaned.

IAQ criteria include a set of air samples, monitoring of human exposure to pollutants, a set of samples at building surfaces, and computer simulation of indoor airflow.

IAQ as part of the Indoor Environment Quality (IEQ), which includes the IAQ as well as other physical and psychological aspects of living indoors (for example: lighting, visual quality, acoustics, and thermal comfort), is the highest air pollution in developing countries and is by far the deadliest global risk. The main source of indoor air pollution in developing countries is the burning of bios (eg wood, charcoal, fertilizer, or crop residues) for heating and cooking. Exposure to high levels of particulate matter has resulted in between 1.5 and 2 million deaths in 2000.

Research has been done on natural displacement over the past three decades. The natural and turbulent flow in the chamber was first studied by Alder and then by Smidogil, and air displacement was used in these experiments.

Providing the comfort of employees in office spaces is necessary. This study shows that air conditioning systems usually do not have a good function for cooling office interiors with open plan. Therefore, the present study was conducted with the aim of providing a simple and practical solution as an intervention in the current state of architecture. The method of this research is a hybrid method due to its interdisciplinary nature. First, after initial observations and selection of the case sample using an experimental method, effective independent variables were identified and dependent variables were measured by digital devices in a random case sample (Yadavaran project office building). Fluent analysis was used to validate the indoor airflow data and using CFD software, their method was analyzed. The results show that the state of the architecture and the placement of its components have a great impact on the quality of indoor air flow and energy consumption of air conditioners. The tests also indicated that the height of partitions in office spaces, the position of the partition walls and the position of air conditioning nozzles should be calculated based on the conditions of each office unit and their correct design improves the quality of air conditioning, provides comfort to residents and also reduces energy consumption.

The development of artificial ventilation systems, in addition to high energy consumption, bring a lot of pollution to the environment. However, different climatic regions of the country due to the geographical location and wind and sunlight have conditions that if used properly, these natural forces can be used to respond to the ventilation of buildings and reduce energy consumption to create comfort conditions.

The present study shows that by taking advantage of valuable experiences and works left by predecessors and updating them according to today's facilities and needs, an important step can be taken in saving valuable energy resources.

In general, the methods discussed in this article are an example of natural ventilation solutions for the building are: one-way and two-way ventilation windcatchers via the proper design and implementation of windows and openings in a floating or chimney-based ventilation building with the ducts such as central stairs, skylights, etc. allow heat transfer outside the building.

Using simple ways such as choosing a suitable cover for flooring around the building, using green space and water, etc. to cool and increase the air flow, using these solutions in the design, in addition to providing comfort conditions reduces energy consumption in buildings as well. In the last decade, the environmental conditions of museum exhibition facilities and storage facilities have been the most important factor in preserving collections and artifacts. Pollution, humidity, and lighting can potentially destroy the crop or even weaken or destroy the material cultural goods that are preserved and displayed in the museum complex.

In the museum environment, temporal stability and spatial integrity of internal microclimatic parameters are essential primarily for the protection of works of art and then for static thermal comfort. In this article, according to the purpose of the article, which is the thermal comfort of visitors and increase the lifespan of antiquities kept in museums. Using library studies and after the necessary investigations and determining the appropriate range for the temperature of museums, the effective factors were identified and the best of them were expressed as necessary suggestions to obtain the goals.

In recent years, some European countries have realized that they can no longer ignore the values of built heritage. Preserving the constructed heritage, the conflict between conservation on the one hand and the use of contemporary methods and on the other hand to improve energy efficiency has become very important. In addition, an individual's understanding of thermal comfort (state of mind, which expresses comfort with the thermal environment and is assessed by mental evaluation) must be considered. Balancing energy efficiency, architectural conservation needs for cultural heritage, and thermal applications is not an easy task.

The main purpose of architectural reconstruction in museum buildings is to preserve these valuable works and pass them on to the future generations. The best way to do this transfer is to adapt the historic buildings to the flow consumption needs and thermal requirements. Doing so will not only extend the useful life of the buildings, but will also allow them to be equally admired by present and future generations.

According to Czop, the museum's main purpose is to collect, preserve, interpret and display cultural, artistic or scientific works for public education. The ultimate goal is to make things longer or more durable because they are not just for today but for future generations. In the last decade, the environmental conditions of museum exhibition facilities and storage places have been the most important factor in preserving collections and artifacts. Providing a safe and comfortable environment for humans is also known as one of the main goals of the museum. The need to observe the status of environmental parameters and changes over time (microclassical factors) is one of the main environmental factors in museums, along with temperature, humidity, pollution, light and human factors. These environmental factors can destroy the material-cultural goods that are preserved, protected, and displayed in museum collections.

HVAC designers and engineers should then work with exhibitors to define a compromise between environmentally friendly requirements.

Temperature is one of the factors that is related to the stability of the works. Temperature increases biological activity. Another destructive factor is the increase in chemical decomposition process. Lowering the temperature, increases the relative humidity and thus increases the degradation. Temperature fluctuations on the other hand cause expansion and contraction in objects, which if the expansion and contraction is sudden and rapid, can cause damage and deformation of objects. This condition is very dangerous and harmful for objects made of composite materials. These temperature fluctuations cause fractures, cracks, and damage to the artifacts. It is also investigated the range of temperature. The high cost of implementing modern methods of good ventilation and the old buildings of many museums in Iran, is one of the main causes of damage to objects in museums.

2.3.10 Factors Influencing the Temperature and Thermal Comfort of the Museum

Temperature in the exhibition areas and storage of collections of cultural figures is the third most important factor of destruction, immediately after humidity and pollution. Temperature is one of the most important factors in the ventilation of museums; as a result, it is the most important factor in saving energy. In the following, the thermal range for objects and humans is discussed. Then, measuring and influencing factors such as air flow velocity are expressed and finally, effective factors such as floor height are examined.

2.3.10.1 Thermal Comfort

Resident Comfort Satisfaction on Indoor Air Quality of Museum building environment in tropical climate and IAQ aspects to be studied as a parameter. Chemical pollutants are gaseous pollutants and comfort factors (temperature, humidity and air movement). A museum is an archive, library, heritage building and gallery that can be considered as cultural heritage and environment. That is, the area in which a person feels emotionally and physically comfortable. Air age and temperature are the most important factors in measuring it.

2.3.10.2 Temperature Range for the Museum

The selected area must necessarily be within the comfort range of the visitors. It is expressed at 23 to 26 degrees Celsius. As a result, the need for temperature balance and minimization of temperature differences in museum exhibition halls is a very important factor. One of the influential factors in that is the environmental climate.

(Table 1). Introduction of mechanical factors

Variable	Type of effects (description)	affecting factors
Age of air, speed and direction of air flow	It has an effect on the direction and type of distribution	Placement of ventilation valves
Age of air, speed and air flow, temperature	At the input temperature to the complex and novelty	External connection valves

| Age of air, speed and direction of air flow, temperature | In energy consumption, efficiency, and so on | Type of air conditioners |

2.3.10.3 Physical Factors

Below, the most important physical factors examined in this article are summarized.

(Table 2). Introduction of physical factors

Variable	Type of effects (description)	affecting factors
Radiant and convective heat	Heat and cold penetrate inside	Material and type of walls
Radiant and convective heat	Heat and cold penetrate inside	Floor height
Air flow speed, temperature through radiation	Orientation in energy absorption	Geographical orientation
Speed and direction of air flow	Building form in the amount of air circulation	Building form
Air flow speed, air flow direction	Way of air distribution, reduction of energy consumption	partition

2.4 Case Studies

According to the studies that have been done, suggestions are presented based on previous studies to improve the temperature conditions in museums:

When constructing museum buildings, the geographical orientation in each area should be observed. It has a huge impact on energy saving. Also, due to the reduction of energy absorbed in the walls, thermal doubts in the building are significantly reduced.

Use insulation on walls and minimize radiant heat from walls and ceilings, use pellets and height, on the floor or insulate it, use of centralized mechanical ventilation in simple forms, use VAV ventilation systems for better air distribution, reduce the number of air outlets in the main halls, adjusting the inputs and outputs of mechanical air conditioners based on the form of the halls, use partitions instead of drywalls. As mentioned, temperature has a great effect on maintaining traces and thermal comfort. And maintaining the temperature of the halls of the museums of antiquities in a certain range and reducing the temperature difference and the so-called isotherm in museums, is one of the most important factors in preserving and maintaining. Therefore, creating an isotherm is essential. In this article, we have tried to give appropriate suggestions for improving the isotherm in the museums of antiquities in order to create more reliable museums for preservation and preservation. Among the results obtained in this study, it can be mentioned that if museums are designed with climatic factors such as geographical direction and floor height, the use of energy-saving systems and distribution systems, suitable air conditioner, and so on; in addition to increasing the lifespan of the works, which is the main goal, it also includes the thermal comfort of the visitors and significant savings.

2.4.1 Central Building of Persian Bank

Cooperation in CFD simulation and evaluation of office building air conditioning system is one of the main concerns in designing air conditioning systems, ensuring uniform distribution of flow and cooling and heating in different parts of the building. This issue is more important in large buildings and the conditions are very complicated in atrium buildings. In the present project, the air conditioning system of central building of Parsian Bank in three interconnected floors and including a design similar to Atrium, was analyzed and simulated using CFD software. The main executor of this project was Mr. Eslami and the group of simulators cooperated in the field of setup, simulation and computing processing. For this purpose, first the geometry of the design, which has many details, was produced in three floors using SolidWorks software. Then, using ANSYS-Meshing software, a suitable computational network was produced for three interconnected floors. The number of computational cells in these three classes was more than 20 million, in which 7 equations of flow, turbulence and energy had to be simulated. The effects of height and natural flows on the building were also considered using the conventional Bozinsky model. Finally, the simulation showed the effects of heat accumulation in the upper floors and suggestions were made to improve ventilation in the building.

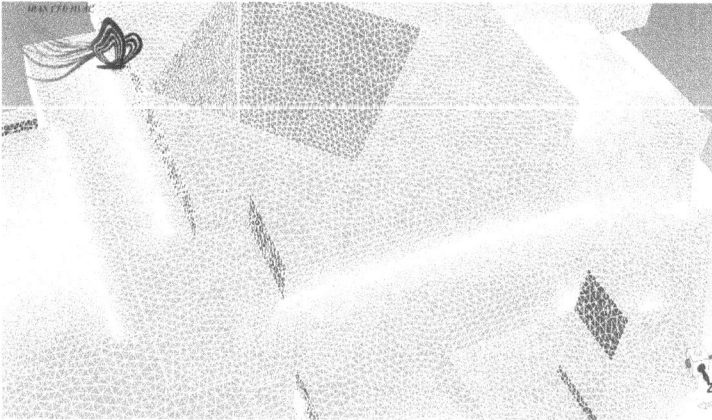

(Figure 44). Simulated example of work done in the complex

(Figure 45). Simulated work done

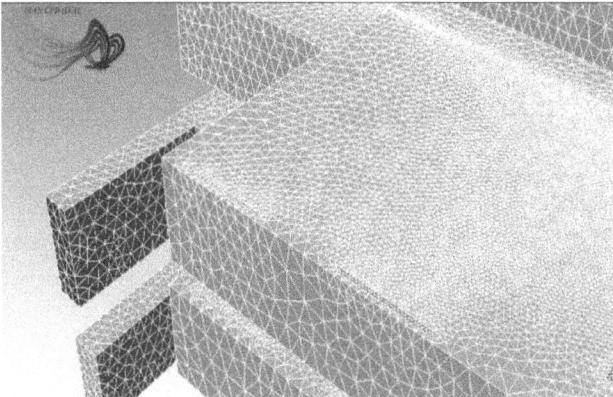

(Figure 46). Simulated from the work done in the complex

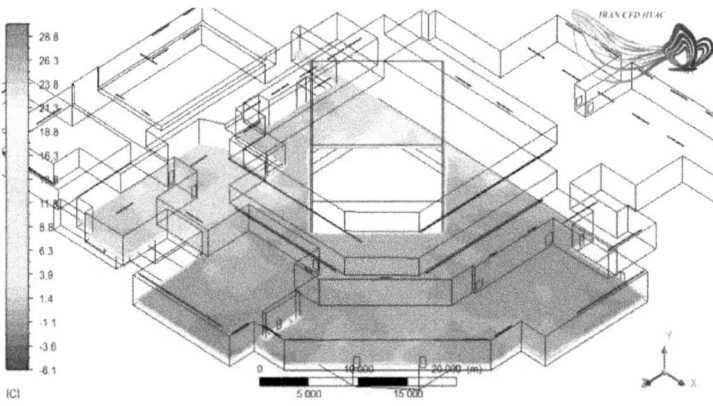

(Figure 47). 3D simulated example of work done in the complex

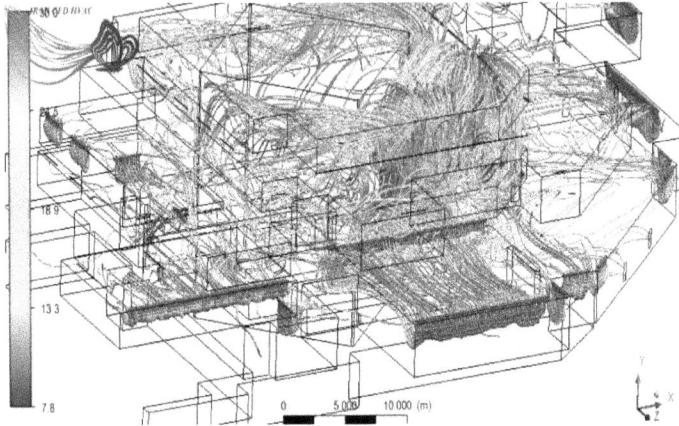

(Figure 48). 3D simulated sample

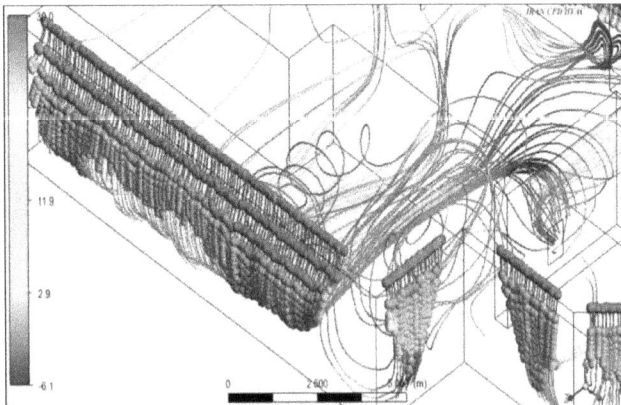

(Figure 49). Simulated sample

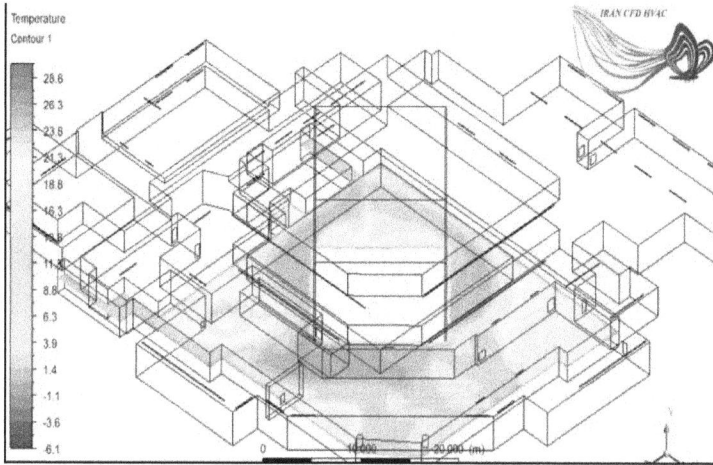

(Figure 50). Simulated temperature of the complex

In large buildings in order for commercial operation with a large staff and long hours of energy consumption, the discussion provides CFD simulation and improved indoor airflow, employee comfort and energy savings. In the areas marked in the photo, it shows the air flow loss paths that have been tried to minimize this flow loss by using FLUENT software.

2.4.2 Royal Danish Academy in Copenhagen

In the field of near-zero energy buildings, there is an argument that the energy-oriented design concept proposed by the regulatory frameworks should be combined with a greater focus on user comfort and enjoyment. Accordingly, the basis of research theory is that designers should take responsibility for understanding heat flows through building components and spaces in a design that conforms to the micro-thermal conditions in the space to allow residents to meet their needs and control their wants. For example, maximizing the benefits of heat from the sun, which stimulates a set of partitions inside to avoid the risk of overheating. Therefore, it is necessary to test the existing simulation software tools with the aim of modeling and visualizing the interior space, the complexity of the thermal environment. This study discusses how to create thermal comfort maps that can be made using them. Computational fluid dynamic simulation method can integrate energy simulation outputs to support architectural quality. Design decisions were used to design strengthening of a small educational building on average radiant temperature maps. Copenhagen, thermal opportunities of movable interior partitions (user-managed) can estimate a new layer of information to the designer. Thermal maps are used in the architectural design process. It is discussed the use of the standard energy simulation convenience output as a reference. The capabilities and limitations of the method are evaluated.

(Figure 51). View of the Royal Danish Academy in Copenhagen

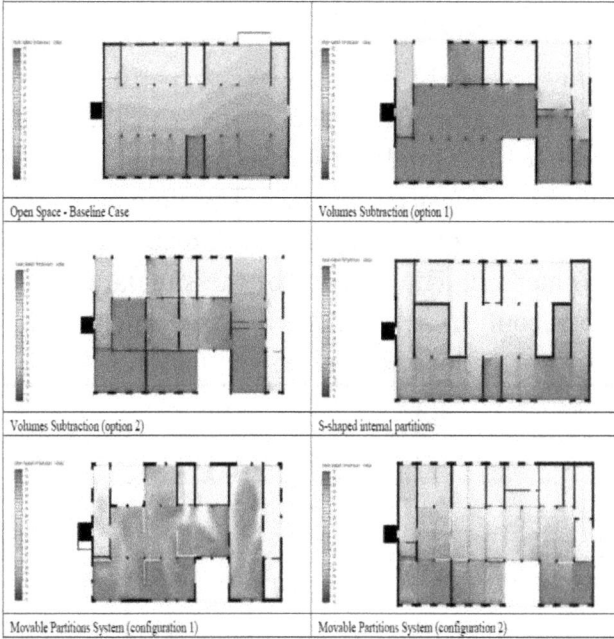

(Figure 52). Complex simulation (movable partition system)

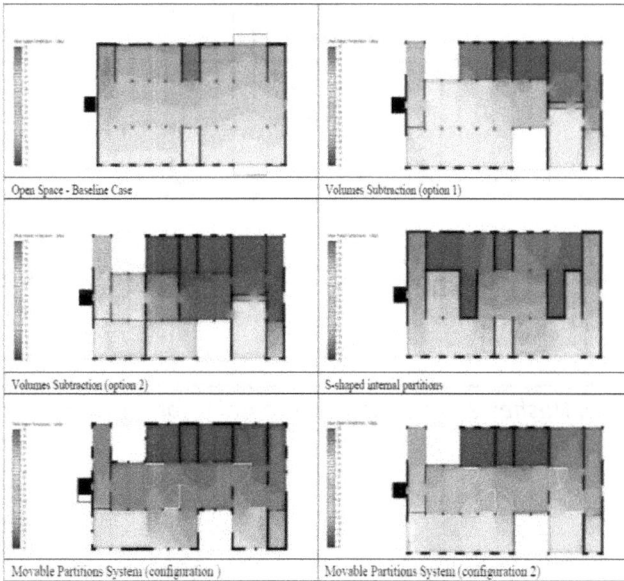

(Figure 53). Standard energy simulation

2.4.3 Data Analysis by CFD Method

In order to accurately evaluate the experimental data obtained from measurements and changes in the field of interior architecture, simulations were performed on a case study of an office industrial complex. The simulation of this research has been done using CFD (Computational Fluid Mechanics) method. The purpose of the simulation in this research is to confirm and analyze the experimental results of in-situ air flow velocimetry using simulation.

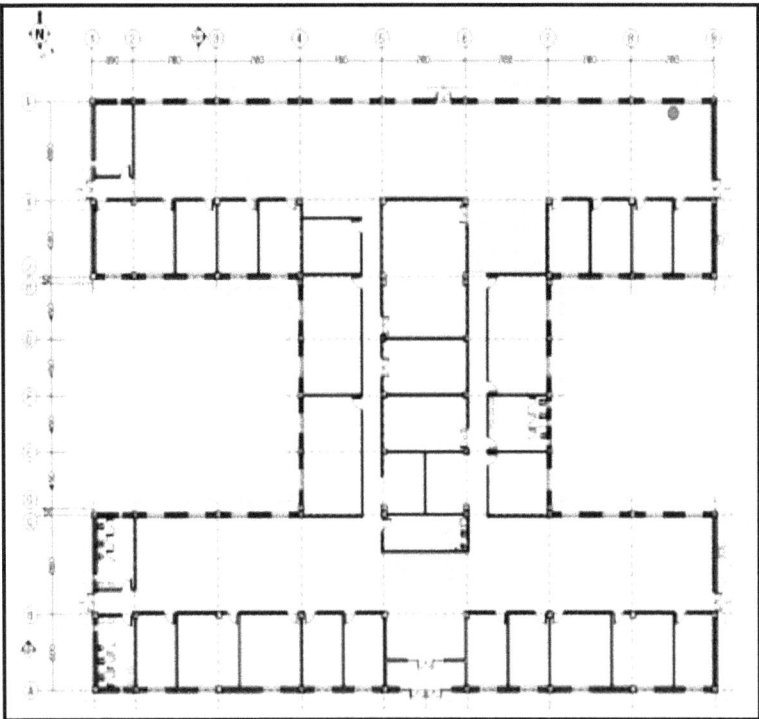

(Figure 54). Hashed space and selected point for measurement tests

(Table 3). Data related to the experimental test of temperature measurement in office building partitioning space before intervention in interior architecture

Sep 16	Sep 9	Sep 3	Aug 28	Aug 20	Aug 9	Jul 30	Jul 24	Jul 18	Jul 11	Jul 7	Jul 1	Test points	Time and steps of temperature measurement tests
27	28	31	28	27/5	27	28	28/6	29	31	32	31	A	The first stage between 10 and 11 o' clock
28	31	29/5	28	29	29	28	29/5	28	29	31	31/5	B	
30	31	30/5	29/5	28	28	30	30/5	29	31	29	32	C	
27	27/5	26	26/5	27/5	27/5	26	27	28	28	27/5	27	A	The second stage between 12 and 13 o' clock
26	26	27/5	28	27	27	27	28/5	26	28	28	28/5	B	
30/5	30	29	31	31	31	29/5	29	30/5	30	29	31	C	
27	26/5	27	26	25	25	26	26/5	27	25/5	25	26	A	The third stage between 14:00 and 15:00
27	28	28/5	26	27/5	27/5	28	27	26	27/5	27	28	B	
29	28	29	30/5	30	30	28/5	28	29	30	30	29/5	C	

Source: Rahai & Roasai, 2015

(Table 4). Data related to the experimental test of temperature measurement in office building partitioning space after intervention in interior architecture

The average temperature of the points before the intervention in the interior architecture

Sep 16	Sep 9	Sep 3	Aug 28	Aug 20	Aug 9	Jul 30	Jul 24	Jul 18	Jul 11	Jul 7	Jul 1	Test points	Time and steps of temperature measurement tests
24	23	2.54	23	2.53	25	24	2.53	23	24	2.55	28	A	The first stage between 11 and
22	23	24	2.54	23	22	23	23	24	2.52	23	22	B	
24/5	26/5	27	25	24	26/5	24	25	25	26	27/5	26	C	
26	26	25	25/5	23	24	25	25	26	24/5	23/5	23	A	The second stage between
22	21	19	19	21	19/5	22/5	22	21	19	23	22	B	
21/5	20	20	21	22/5	21	20	20	21/5	22	21	20	C	
26	26/5	27	26	25	27	26/5	25	25/5	24	26	27	A	The third stage between 15:00
23	22	23/5	24	24/5	23	22	23/5	21	22/5	24	23	B	
20	21/5	21	20	20	19/5	18	19	20/5	19	21	20	C	

Source: Rahai & Roasai, 2015

For example, the area selected to simulate the hall of the northern part of the office building is in the figure. First, for geometric modeling and networking of the area plan in which the experimental tests were performed, Gambit software was used and its boundary conditions were determined (airflow inlet and outlet valves, position and height of partitions were installed on 3D grids). Then, the grided model for numerical calculations and simulations related to temperature and air flow were analyzed using Fluent software and by the data obtained from experimental tests in (Table 3) and (Table 4). The simulation results show that the current situation is consistent with the results of experimental temperature measurement tests at the site. It turns out that the simulation of this research has the necessary validity. Based on this, the simulations are presented as static pressure difference temperature contours in figures 55 and 56. (Rahai & Roasai, 2015).

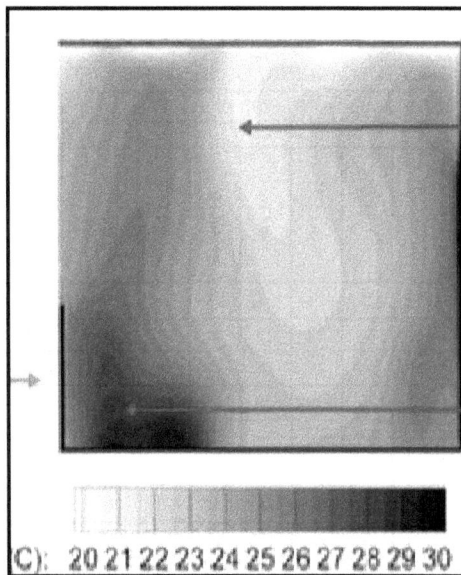

C): 20 21 22 23 24 25 26 27 28 29 30

(Figure 55). Contour positive and negative pressure differences before changes in interior architecture

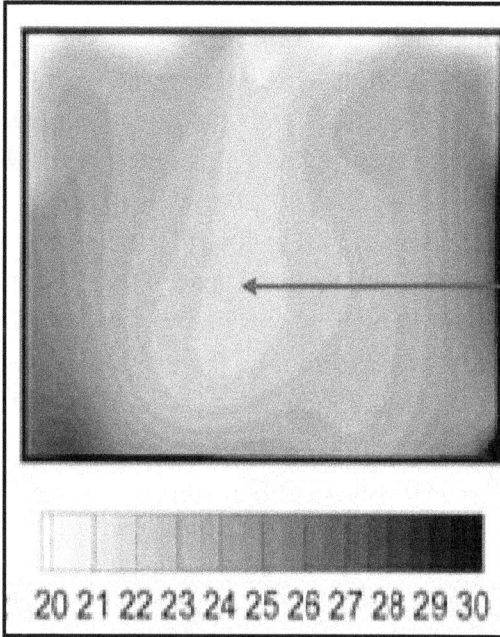

(Figure 56). Contour positive and negative pressure difference after changes in interior architecture

As shown in the figure, the temperature status in the upper middle part of the positive pressure zone (light part) has more favorable conditions than the lower level in the low pressure zone (bold part) of the area where employees are sitting in the partition space. This difference between positive and negative pressure indicates that the airflow between the partitions, especially the low level, is not the same and the thermal comfort conditions of the staff at the low levels are not favorable. This condition has arisen in a situation in which the partitions have been executed with a height of 1.5 meters and the space above the partition is open to the dry ceiling. This factor causes a large part of the air flow out of the air diffuser vents to escape from the open space between the partitions to the ceiling and not to be transferred to the lower levels caused the temperature to rise at a low level. Also, in order to observe the changes in temperature after the intervention in the interior architecture, a static pressure difference diagram was presented. This diagram shows that the temperature conditions have decreased at a low level (positive pressure zone). Also, the change in the height of the partition walls (partitions) from 1.5 meters to 3 meters caused to control most of the air flow that was transferred from the open space above the partition and the dry ceiling to other spaces before the changes in the interior architecture. Accordingly, the air flow between the partition space and the range of staffs' activity should be established in the same way (figures 57 and 58). In order to confirm the results of this research, a survey (in the form of interview) was conducted on the employees of the office building, and many were satisfied with the changes made. It was said that this method can be considered as an executive solution to solve problems related to ventilation.

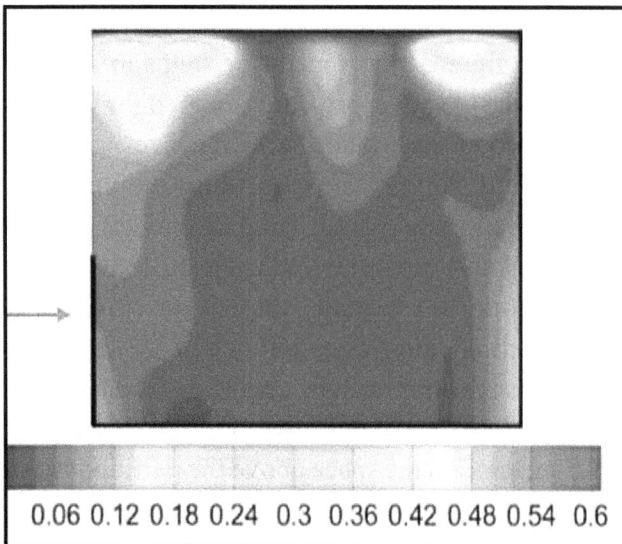

0.06 0.12 0.18 0.24 0.3 0.36 0.42 0.48 0.54 0.6

(Figure 57). Airflow contour before intervention

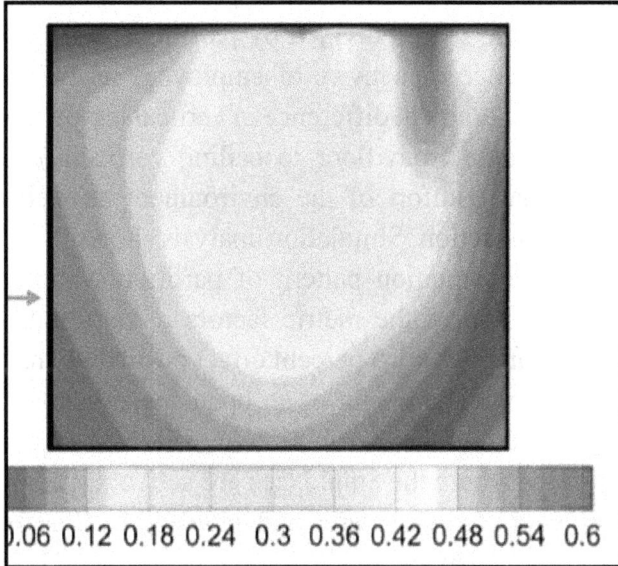

.06 0.12 0.18 0.24 0.3 0.36 0.42 0.48 0.54 0.6

(Figure 58). Air flow contour after intervention

The result of this analysis is that the provision of adequate air flow in indoor spaces using air conditioning systems depends on various factors such as the location of airflow vents, dimensions and size of spaces, and the shape and location of physical elements. The use of air conditioning systems with optimal performance has a significant effect on reducing the indoor temperature of buildings, especially in areas with hot weather conditions. Considering that the effect of optimal performance of air conditioning systems on health and improving the thermal comfort of employees and saving energy consumption plays an important role, this study examined the quality of air conditioning in office spaces using architectural interventions. And in order to achieve more accurate results using the CFD method after its validation with experimental findings, has analyzed the condition of indoor air flow. The results of the case study showed that the lack of uniform distribution of air flow in the office building and the inefficiency of air conditioning systems to cool the spaces divided by partitions, especially at low levels, (range of activity of employees in a sitting position) caused a gradient difference in vertical temperature (temperature change from floor to ceiling vertically) and disturbed thermal comfort of the environment as well as employee dissatisfaction. Simulation analysis shows that the design and implementation pattern of partition walls and airflow vents is one of the metric factors of partitions in terms of height and has a 1.5 percent effect on airflow inside the office building. The height of an open space below the dry ceiling has caused the distribution of airflow among the partitioning area where the employees are active, especially in the lower levels, to be scattered and undesirable. Based on the architectural changes made in the experimental and simulation tests between the partitions and around the airflow vents, the results were obtained by changing the height of the partition plates to increase in meters and distribution of 3 m to 1/5 of the simulations is shown, if the wall height of the partitions is provided from a few meters above the ground. In addition, these changes make it possible to control the gradient of air flow in the vertical temperature level by using the air flow provided in the area where the employees are working. This will provide them with thermal comfort and reduce the energy consumption that was increased due to the greater operation of the air conditioners to cool the interior. In addition, this method saves energy consumption, which is economically important. However, in addition to considering the executive solution that was examined in this study, it is possible to improve significantly the quality conditions of air conditioning in interior spaces of the buildings by studying more carefully on other physical elements that form spaces and paying attention to their dimensions and location in the design.

2.5 Conclusion

In general, the design of a museum is an important issue because it is one of the national values of each country. Due to the existence of historical and precious objects in museums, their preservation is an important issue in the long run. For this purpose, improving the indoor air flow and isotherm is very important because the lack of isotherm inside the museum building causes damage to these objects and facilities. In this chapter, by introducing the necessary elements for designing a museum and the topic of CFD Method, we have tried to point out the theoretical foundations necessary to improve the air flow inside the museum.

Chapter three:

Simulation with FLUENT

3.1 Introduction

This study was conducted with the aim of improving the quality of air conditioning in the interior of various buildings and providing comfort for people, which can be achieved by intervening in the architecture of the current situationn. Since many construction studies are interdisciplinary and require special hybrid methods, the method is the CFD type. Because the purpose of this study is to improve the air flow status with the help of architectural variables in order to improve the quality of natural ventilation and prevent damage to the building and the health of the people and in the form of interdisciplinary architectural research (mechanics and architecture), a special combination method is needed. Indoor air quality is related to the air quality inside and around the building, and to the health and comfort of the building's occupants. It also depends on gases such as carbon monoxide, radon and volatile organic compounds, microbial contaminants (molds - bacteria) or any stressful mass or energy to be able to create unfavorable conditions and affect health. Pollution control, treatment, and the use of ventilation to reduce pollutants is the main method for improving indoor air quality in most buildings. In a residential unit, we can improve the air quality inside the building with very simple solutions, such as using the fresh air supply system or regular cleaning of home appliances.

3.2 Simulation Process

In this research, which is the result of the authors' experiences on museums, first the building of Ahmad Shahi Palace complex was selected because it is a historical museum and the quality of temperature is of special importance. Then, by preparing a map using simulation and software (FLUENT[1]) and (CFD) and case studies, volume modeling with Auto Cad program, simulation was performed. Simulations in this study were performed by CFD (computational fluid dynamics) method and Gambit and Fulent programs. In this way, the space in question was first networked by Gambit software. Then the boundary conditions were defined and thus, numerical calculations and final simulation with Fulent and validation of CFD method in this study were performed with experimental tests and matching their results. Finally, the results were presented based on the status of architectural variables. In general, in this research; study methods, field survey, observation, experimental tests and simulation methods have been used. IAQ criteria include a set of air samples, monitoring of human exposure to pollutants, a set of samples at building surfaces, and computer simulations of indoor airflow.

With this measure, air quality in centers such as museums is easier and more accessible. This method is used to help better preserve antiquities. The prototype was developed and cultured in the laboratory of the Norwegian Institute for Air Research. Factors such as light, temperature and humidity have always been measured in museums since ancient times and this factor has been the formation of this research.

Ellen Dahlin, archaeologist, says "many museums do not have facilities for measuring air pollution. This may be that they do not have the right tools to do so. Also, these instruments and devices for measuring air pollution are very expensive. And that is why they are unaware of these pollutants". One of the advantages of using this measuring

---

[1] FLUENT Software is a computer-aided engineering software in Computational Fluid Dynamics (CFD) for modeling fluid flow and heat transfer in complex geometries. This software allows complete network change and flow analysis with unstructured networks for complex geometries.

device is that it allows users to read the result on the website. Previously, analysts had to consider using two different sensors in different laboratories to analyze data.

Trege Grunft, chemist, says "the latest point about this measuring device is that it is now more evolved in two parts than its predecessor. First, it is more sensitive to acids in space, and second, it is more sensitive to the amount of pollution and measuring and determining the type of pollution, all of which can be detected with this small device. Operators of this device can view the result on their computer screen. Green light means a clean environment away from pollution and yellow light means the possibility of pollution and red is a sign of pollution and danger to the environment.

Other partners in this study project investigate the effects of increasing the effectiveness of this method to develop the protection of cultural and historical monuments and increase the air quality inside museums.

By collecting data related to the proposed method, the data is implemented and simulated using Fluent software. In this research, the computational fluid mechanics simulation strategy is used. And by CFD programs, these softwares were used according to the results obtained from the literature. In this part, the volume was networked using Fluent and with the help of Gambit program, the architecture of the specified area was analyzed using the data analysis method.

3.3 Fluid Dynamic Analysis

The spatial microclimate distribution in the building has been evaluated and varies on the type of air dispersion through CFD analysis. Therefore, the CFD boundary conditions are provided by the BEPS program. And the mass-energy balance equations are the same for all internal applications, while the boundary conditions change for each case. The diffuser uses a kinematic method, which indicates that the diffuser acts like a valve characterized by geometric dimensions, output and kinetic instability. These outlet boundary conditions are specified to determine the air circulation rate from the extraction of 4 grills. Internal thermal loads are allocated using individual models and lighting facilities. The Fluent Airbag program, especially the zero equation turbulence model has been used, is a valid approach in numerical solution methods, indoor air flow and less mobile (in the case of calculation) than the standard k-approach. Sensitive analysis has shown that these results are independent of the network structure. Considering only the combination of air conditioning diffusion strategy, the following supply equipment is modeled.

Ceiling-mounted diffusers and wall-mounted grilles have fixed vans due to software limitations, while for other emission equipment, the deflection angle is below horizontal in summer and below vertical in winter. Air is extracted from the building using 4 grills located at the bottom (near the floor) in the corners. Any CFD simulation must guarantee temperature and relative humidity. Adjust point values measured with a dual sensor mounted on the volume of preferred artwork. This system simulates real automatic control. Therefore, the required calorimetric conditions of the air depend not only on the thermal loads but also on the type of diffuser. These conditions include thermal performance, hygrometry, and kinetic fields in the building.

3.3.1 CFD Check of Museum Air Velocity Distribution

In this study, Computational Fluid Dynamics (CFD) investigates the distribution of airflow within the museum, which is one of the most important historical sites in Tehran.

The equations of stability, motion and energy are solved together with the K-e turbulence model using a commercial CFD package. Stable and non-compressive flow hypotheses have been developed to simplify the simulation and reduce the computation time. Samples of speed in different parts of the ground floor of museum, circulation zones, and whirlpools in some areas have been shown to indicate poor weather in. Also, low speed areas are noticeable in some places, especially behind the columns and any obstruction of flow. This study concludes that the distribution of natural airflow can be improved through a different combination of open windows and doors based on wind speed and direction. Chung and Hesu (2010) examined ventilation efficiency of different natural ventilation patterns arranged by two inputs and two outputs at different locations with a full-scale test chamber. The results show that the locations of the window openings, strongly affect the natural ventilation of the building efficiency. Air flow distribution, carbon dioxide concentration and air change rate for different window open angles and inlet air velocity are investigated. The simulated and experimental results show the typical ventilation characteristics of an axial center window. It also allows designers to specify the right window opening angle for different outdoor air velocities.

Cho (2008) performed in his work, all studies using CFD to simulate the distribution of airflow inside the museum in stable conditions with respect to windows and doors that are open during the visit. Humidity is not considered the distribution of emitted gases, which is no less important than the distribution of air.

3.4 Survey of Ahmad Shahi Museum Palace

This mansion was built in the late Qajar period for Ahmad Shah summer dormitory in the middle of Niavaran garden with an area of 800 meters on two floors with a gabled roof. One of the main features of this building is the decorations and brick facade used throughout its exterior. The bricks are of molded pattern type with various designs of buffy in color. The entrance of the building is located on the south side, which leads to the pavilion by several stairs along the oval pool covered with tiles.

Ahmad Shahi mansion in the second Pahlavi period was renovated and new additions were made to it and its interior furniture was completely changed to be used as a place of work and residence of Reza Pahlavi. The ground floor of this mansion includes a hall with a marble pool in the middle and 6 rooms and 2 corridors around it. Decorative objects made of silver, bronze, ivory, wood, gifts from different countries such as India, paintings and goblins and signs and medals are displayed in this space. Also decorative mineral objects and stones, moonstone, and several plant and animal fossils, are demonstrated. The second floor of the mansion consists of a central hall and a four-sided porch. On all four sides of the central hall, which was used as a music room, there is a wooden shelf with a showcase. The porch is surrounded by 6 columns with a square cross section with a brick facade and 26 circular columns with a gypsum facade. The lion and the sun on the bed can be seen on the forehead of the wall on the north side of the porch. After the Revolution, during the protection and restoration of this building, the lower part of its walls was repaired and in May 2000, at the same time with the week of cultural heritage, this mansion was opened.

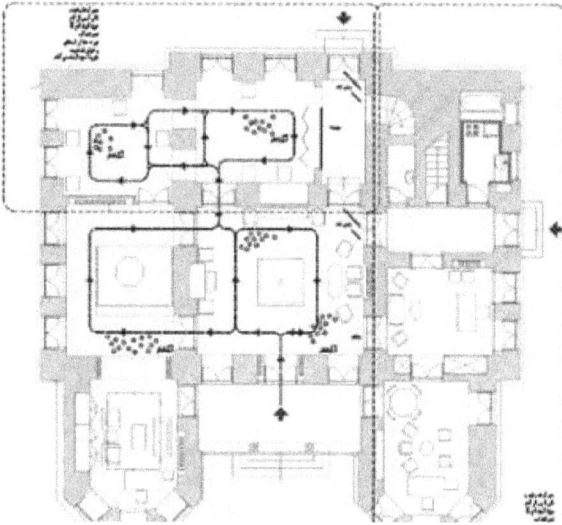

(Figure 1). Plan of Ahmad Shahi Palace

(Figure 2). View of Ahmad Shahi Palace

(Figure 3). View of Ahmad Shahi Palace

(Figure 4). Plan of Ahmad Shahi Palace

(Figure 5). 3D view of Ahmad Shahi Palace

(Figure 6). Plan of Ahmad Shahi Palace

(Figure 7). Plan of Ahmad Shahi Palace

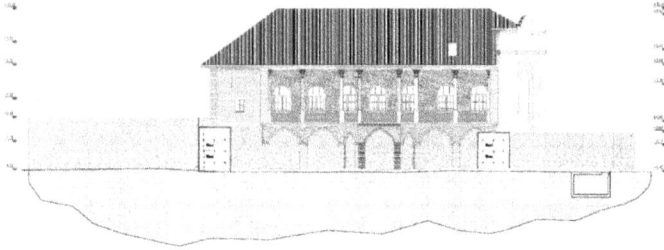

(Figure 8). View of Ahmad Shahi Museum Palace

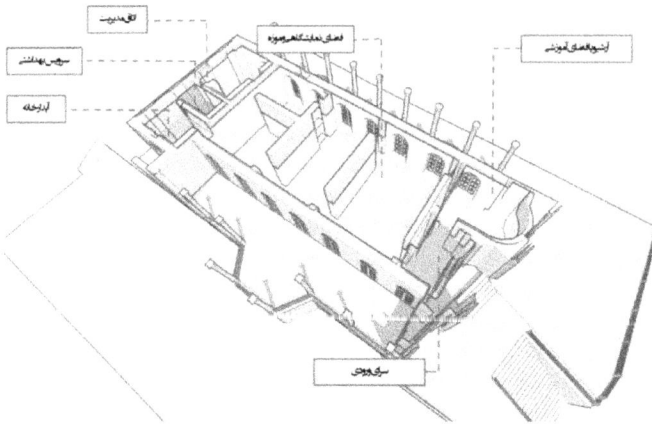

(Figure 9). Three-dimensional view of the palace

3.5 CFD Analysis

3.5.1 External CFD Analysis

Exterior CFD analysis of the building explains the speed and air pressure around the building due to wind. This information is used to assess the comfort of passers-by, calculate the exact pressure coefficient to simulate natural ventilation, energy plus local pressure to determine the location of outlets and chimneys, etc.

3.5.2 Internal CFD Analysis

The internal CFD analysis of the building provides the distribution of temperature and speed and air pressure and several other parameters throughout the interior of the building. It also calculates the air life parameter, which indicates the novelty and freshness of the air in the space and is a comfort index. This information is used to assess the impact of different types of air conditioning systems and different natural ventilation schemes as well as to assess indoor comfort conditions.

3.6 Simulation of the Current Situation

Check different scenarios

The airflow inside the museum was tested under different conditions. These conditions were analyzed in the form of two- and three-dimensional models. In these analyzes, as much as possible, all available cases were analyzed according to criteria and standards. Before the intervention, different conditions were first analyzed as follows:

General 2D modes

a- The first model: longitudinal

The bottom of the model in longitudinal section and in the picture below, all the modes of indoor air circulation in different positions were examined. These modes are as follows:

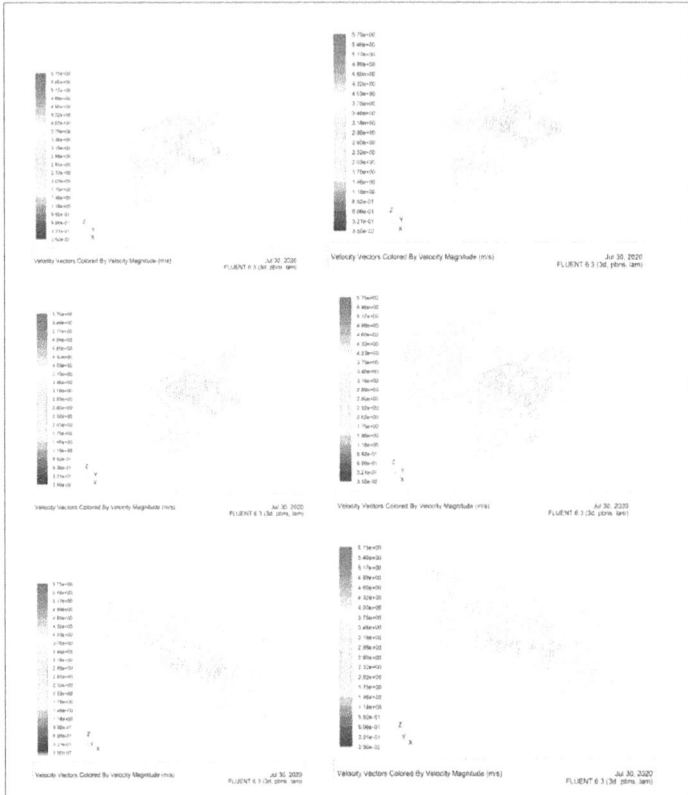

(Figure 11). Longitudinal vector diagrams of air velocity, source of the author, 2020

b- The second model: cross-sectional

The model was examined cross-sectionally and all indoor air circulation conditions in different positions were investigated in Figure 12. These cases are as follows:

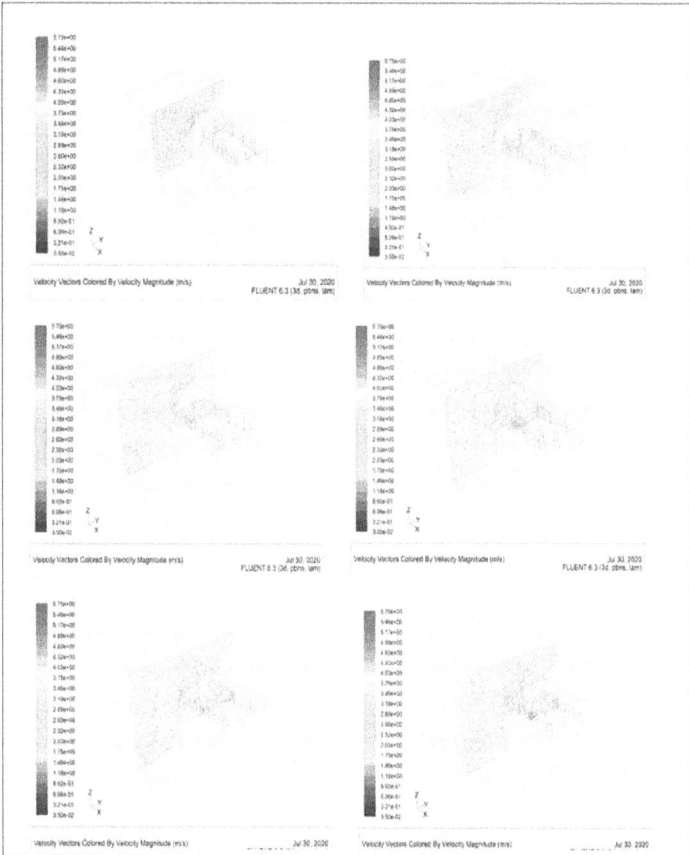

(Figure 12). Cross-sectional vector diagrams of air velocity,
source of the author, 2020

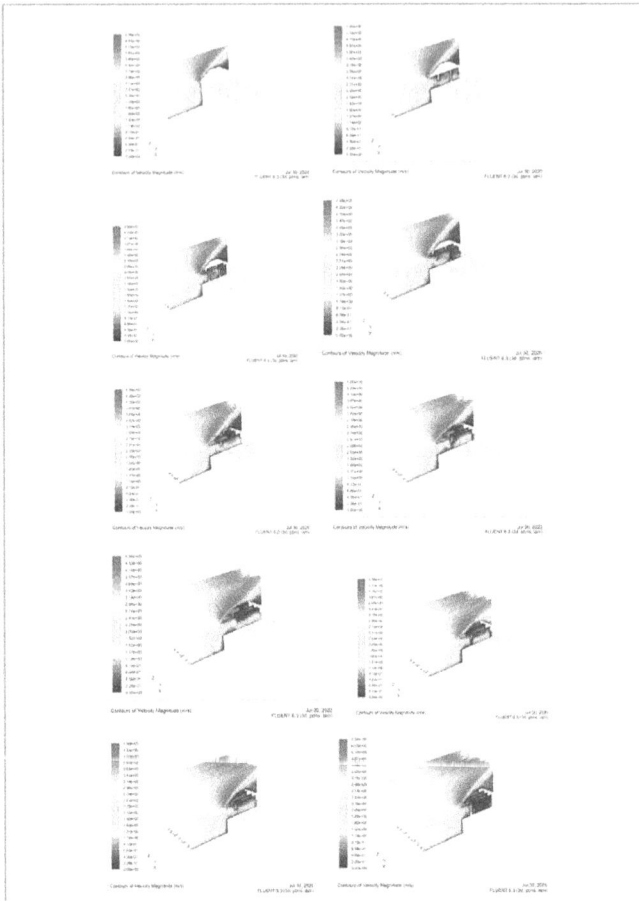

(Figure 13). Diagram of static pressure contour, source of the author, 2020

In the pictures below, the three-dimensional model of the museum was analyzed. According to the picture, it was considered that with the opening and closing and changing the suction and blowing conditions of blowers and suckers, several states were assumed. Based on the results of two-dimensional analysis, a mass of the interior space of the shed in accordance with the following images was assumed to be full space (office floor, warehouse, or empty section), which in some cases was applied as follows. These conditions were investigated in different models:

Indoor air speed with 2 partitions

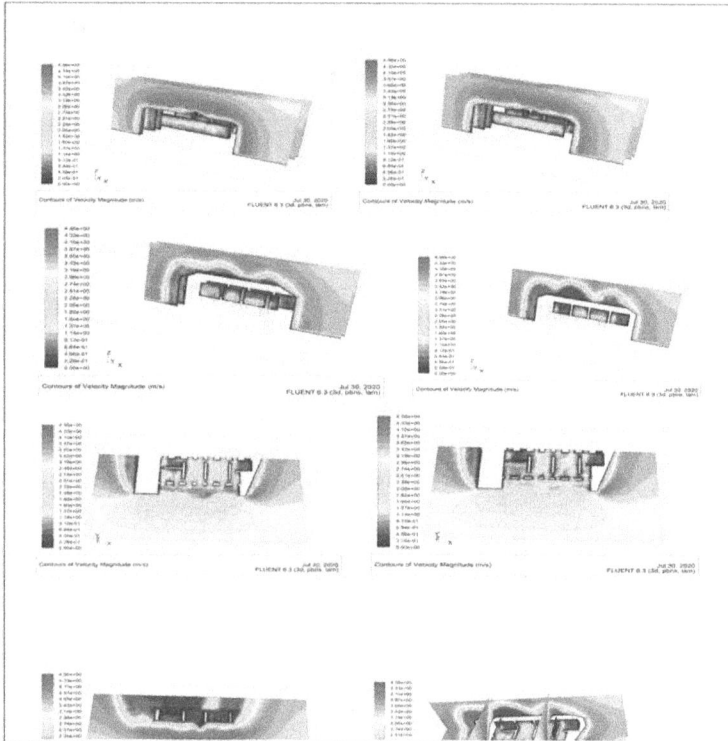

(Figure 14). Two back doors open to the right wind, two doors facing the left wind blower

The model in longitudinal section and all indoor air circulation conditions were examined in different positions. The first model is the indoor air velocity with 2 partitions, the modeling of which is as follows:

Indoor air speed with 2 partitions:

In this model, the indoor air velocity is modeled with two internal partitions, as shown in Figure 15.

Two back doors open to the right wind and two doors facing the left wind blower

In Figures 15 and 16, the two back doors to the right wind were opened and the two doors to the left wind blower were modeled.

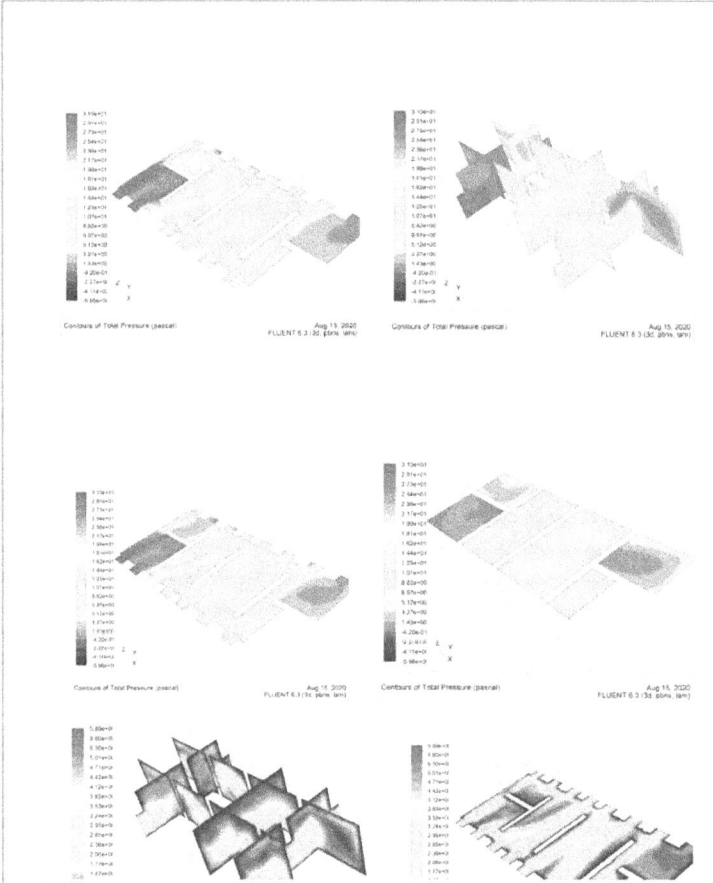

(Figure 15). Two back doors open to the right wind and two doors facing the left wind blower, source of the author, 2020

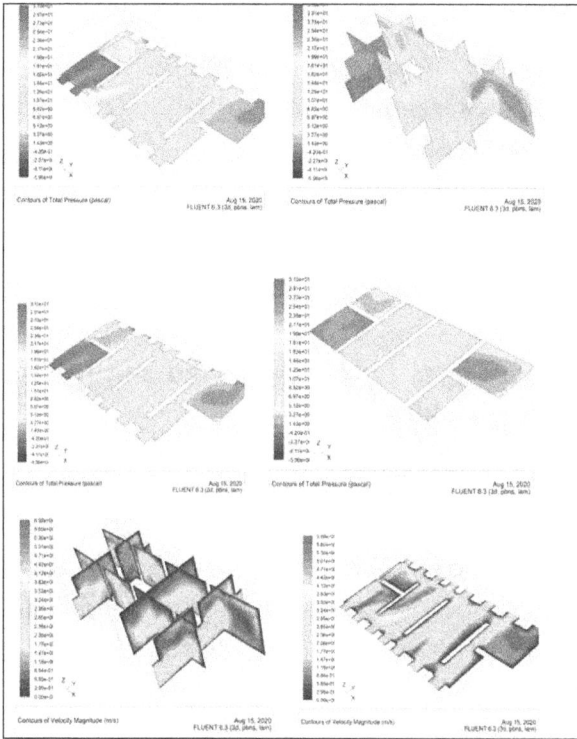

(Figure 16). Two back doors open to the right wind and two doors facing the left wind blower, source of the author, 2020

3.7 Conclusion

Indoor air speed with two partitions:

By examining the cross-sectional velocity contour, we find that the partitions prevent the free flow of air and also reduce the velocity near the walls, which according to the law of mass conservation leads to an increase in velocity in other places. The maximum speed in this case is about 4.5 meters per second, which seems like a lot for a museum and human residence (3 meters per second maximum allowed).

Examining the top view of the speed contours, we conclude that the presence of two partitions creates a low-velocity area in the center and top of the shape and will increase the speed around. In this case, the speed in the middle is 1.5 meters per second, which seems appropriate. At the front, the speed is about 3 meters per second.

So we can conclude that this arrangement will reduce the speed in the museum area and the number of visits, will increase the speed in the service and library area.

Two back doors open to the right wind and two doors facing the left wind blower

In this case, we see that the combination of doors and blowers creates a low pressure area on the left and a high pressure area on the left, which can be predicted due to the combination of open and closed doors. In addition, the maximum speed is observed near the blower, which is about 5.9 meters per second. It is observed that the pressure in the cross section decreases to move from left to right. In addition, it is observed that the visiting room is located at an average speed of about 3 meters per second and the speed decreases on both sides (left and right). Also in this case, the air flows effectively in different parts and can not just enter the training part on the right.

Two doors facing each other randomly in the middle room in the form of a blower and an opening exit

When two doors are randomly facing each other, a kind of short circuit of air is created inside the room in such a way that the incoming air from one door without circulating in the room leaves the next door. As a result, it will not have much effect on the average speed of air flow inside the room. Examination of velocity vectors and pressure as well as velocity contours also confirms this matter. The pressure distribution inside the room is almost constant. Also, since most of the air comes out of a window directly from the front window after entering, only the speed in this part increases. However, the entry of air into the training area will also increase the speed in this area to some extent. As a result, it can be said that this is not an optimal arrangement for the room.

Two blowers and three open windows

In this case, the pressure contours show that the air pressure on the left (service section) was higher and decreases by passing to the right (training part). Speed contours also show that the arrangement of suction and windows causes the air in the room to flow from left to right, which will have a positive effect on the air quality of the room. Due to the fact that one of these open windows is located on the opposite side of the blower, most of the air flow comes out of the window directly. In other cases, the combination of blowers and windows will speed up the air flow. The air speed in the middle part is about 1 and in the other two parts it reaches up to 2 meters per second.

Three blowers and two open windows

In this arrangement, the presence of blowers creates two high pressure areas on the right and middle. Also, due to the fact that the open window is in front of the air blower, the air comes out directly from. The air contours also show that the speed is high in this area and moderate in other places. In this case, the air comes out of the blower of the window after passing through the left side and the right air comes out of the three blowers. As a result, the speed is low in some parts of the room and is not effectively ventilated.

A blower and an open window on the opposite side

In this case, a blower is located on one side of the room and there is an open window in the next part. Looking at the pressure gauges, we notice that the pressure is more on the left and is moderate in the visit section. As expected, it is low due to the window on the right. Looking at the speed contours indicate that in this case, the air flows effectively in the room and ventilates it. But the problem with this arrangement is that air cannot enter the area to the right of the museum. As a result, a low-velocity zone is formed in this area that is not well ventilated.

Two blowers and two open valves

In this case, the air circulates effectively in all the middle parts and cools it, but two parts are observed at low speed on the left and right. In this case, the pressure on the left side is maximum and reaches its minimum in the training area. The pressure is also moderate in the middle. By changing the position of the blowers, the flow pattern in the middle chamber and its surroundings is changed and the velocity distribution is uniform. It seems to be the optimal speed distribution arrangement for room ventilation.

In conclusion, it can be said that the first step in designing exhibition spaces based on stabilizing the internal temperature of the museum is to have a clear mind of what you want to show, how many subjects, the planned subject to be shown in the year, how often they will change, what types of circular exhibitions do you have in your schedule, if you have a permanent collection how many parts of it will be permanently displayed, do you display most large-scale topics or small sections, 3D subjects are displayed, and fragile photographs or old prints or drawings and designs are placed in boxes or on the base of columns. Depending on the type of elements in the museum, their location can be stabilized so that the museum temperature remains the same. With strong designs and plans, you can also decide on the degree of flexibility needed and the specifications of your galleries and the size and environmental quality of the spaces. There are few fixed rules in this section and the guidelines are usually general.

Flexibility is key issue: is it possible to build the gallery space based on the type of natural ventilation? Or do we really need mechanical tools? Do you have effective facilities for deploying video devices?

The exhibition space of museums is a very sensitive. It is an important part of this collection that its characteristics affect the collection as a whole. An exhibition is a special type of space in which, in addition to the human-space relationship, there is also an integrated relationship between space and the object. In parts of the exhibition that have fixed exhibition sets, the architecture can be adapted to objects as much as possible, but in flexible parts, this is only possible through a lot of decorations and arrangements.

(Figure 17). Picture of the exhibition hall

What is most important in an exhibition hall is to present the best works of art. Despite the advanced technology of artificial lighting, natural ventilation and assistance can still be useful in the continuation of works of art. Through special elements, the air passage can be moved and the air can be circulated as desired.

References:

Persian references

1. Akbar Beigi, Mahsa. (2013). Design of the outdoor furniture collection of the palace museum garden with a comprehensive design approach. Master Thesis.
2. Akbari, Moeen. (2016). Utilizing the principles of naturalism in the design of the Museum of Natural History of Ilam. Master Thesis.
3. Aghamiri, Hossain. (2008). History of Museum Architecture and International Road and Construction Monthly.
4. Pour Ali, Mostafa. (2011). Phenomenology in Architecture, Quarterly Journal of Research.
5. Pour Fallah, Daei Niaki, Seyyed Omid, Zare Ghadi, Aryian (2018). Simulation of air conditioning system with geothermal energy with solar collector. Amirkabir Journal of Mechanical Engineering.
6. Tommy Cleon, translators: Mohammad Reza Lillian, Mahdieh Abedi, Aryian Amirkhani, and Mansoureh Tahbaz. "Natural ventilation in buildings, architectural concepts, supplies and facilities". Specialized publisher of architecture and urban planning. First Edition, 2010. Pp: 14-60.
7. Jodat, Mohammad Reza. (2002). Museum architecture. Architecture of our magazine. No. 9.
8. Jokar, Farnoosh. (2011). Development plan of Fars Museum of Carpets and Handicrafts. Master Thesis. Shiraz College of Art and Architecture.
9. Jahandideh, Mehrdad & Bahrpeyma, Abd Al-Hamid. (2018). Designing a settlement in Chabahar city based on CFD and creating natural ventilation in its climate. Fourth International Conference on Civil Architecture and Urban Planning at the Beginning of the Third Millennium, Tehran, Permanent Secretariat of the Conference, East Azarbaijan Province Architecture and Urban Planning Association, Tabriz University, Tabriz University of Arts, Alborz Architecture and

Urban Planning Association, Alborz University, Netaco Study Consortium, Pai City Building Institute.

10. Jahandideh, Mehrdad & Bahrpeyma, Abd Al-Hamid. (2018). Investigation of natural ventilation in Deloushi house (Chabahar) using calculations. Fourth International Conference on Civil Architecture and Urban Planning at the Beginning of the Third Millennium, Tehran, Permanent Secretariat of the Conference, East Azarbaijan Province Architecture and Urban Planning Association, Tabriz University, Tabriz University of Arts, Alborz Architecture and Urban Planning Association, Alborz University, Netaco Study Consortium, Pai City Building Institute.

11. Cheghelvand, Mohammad. Kooshki, Reza (2016). Museums and identities. Anthropological analysis of museum visits. Anthropology Letter. 13th year.

12. Hezbei, Morteza; Adib, Zahra, & Nasr Alahi, Farshad (2014). Natural ventilation in Shavadons of Dezful city using CFD modeling. Bagh Nazar Montly 11 (30).

13. Hasan Beigi, Mohammad., & Jiraei, Mahdi (2018). Investigating the role of Sultanabad state merchants and associations in the period of the Constitutional Revolution (1904-1925). Journal of Social and Economic History, (1)7. 1-20.

14. Hosseini, Ilham. (2015). Investigating the role of information technology in museums and science centers in tourism development. Conference on Tourism and Geography.

15. Hokmi, Mehrdad, Akhtarkavan, Mehdi., Satari Sarbangholi, Hassan (2015). Exploring the concept of museum for architectural design Museum International Conference on Architecture, Urbanism, Civil Engineering, Art and Environment. Future horizons, retrospect (look to past).

16. Khoda Karami, Rad, Rashidfar. (2018). Practical methods to reduce energy consumption in a residential complex with a focus approach on the roof of the building. Quarterly Journal of Environmental Science and Technology.

17. Khosravi, Moloud. (2015). Investigating the effective features in improving the quality and quantity of museum architecture

design. 3rd International Congress of Civil Engineering, Architecture and Urban Development, Tehran, Permanent Secretariat of the International Congress of Civil Engineering, Architecture and Urban Development, Shahid Beheshti University.

18. Khajavi, Faezeh., Khajavie, Azadeh., & Mir-Saeedi, Leila (2017). Application of modern wind tower in contemporary architecture and its ventilation analysis, Fifth International Congress of Civil Engineering, Architecture and Urban Development, Tehran, Permanent Secretariat of the Conference.

19. Daadras, Bahareh., & Kazempour, Mahdi (2017). Museum design with a traditional architectural approach. Fourth International Conference on New Technologies in Civil Engineering, Architecture and Urban Planning, Tehran, Salehan University.

20. Raja, Amir. Raja, Omid. (2016). Applications of CFD in urban architecture, metropolitan air pollution control, and energy optimization in buildings. International Conference on Civil Engineering and Architecture.

21. Rigiladz, Parisa., & Faraj Alahi Rad, Amir. (2017). Investigating the intended conditions in the design of the Anthropological Museum. The first national conference of modern research in Iran and the world in Management, Economics, Accounting and Humanities, Shiraz, Shushtar University of Applied Sciences.

22. Raessi, Ilham. (2016). Designing an energy museum park with the approach of optimizing energy consumption in Shiraz, Master Thesis.

23. Zahedi, Mohammad., Haji Ha, Bajareh., & Khayam Bashi, Maryam. (2008). Museum, museumkeeping, and museums.

23. Shoaei, Hamid Reza., & Ahmadi, Naser. (2016). Design of Gorgan Archaeological Museum for sustainable architecture. Master Thesis, Non-governmental, Islamic Azad University, Islamic Azad University of Shahroud, College of Arts and Architecture.

25. Shamse Alam, Razieh (2015). Studying Fars Museum of Handicrafts with the approach of reviving the art of localism. Master Thesis.
26. Shaykh Al-Islami, Hamid. (1999). The importance of recognizing cultural values. Journal of Culture and Life.
27. Shirazi, Bagher. (2000). The appearance of the teachings of the country. Journal of Museums. No. 12.
28. Talebian, Nima., Atashi, Mehdi., & Nabi Zadeh, Sima. (2013). Kasra Museum of Architecture and Urban Planning. Herfeh Publication, the first Herfeh.
29. Araghi, Mohammad. (2012). Simulation and optimization of air conditioning heat transfer distribution in the amphitheater by CFD method. Master Thesis.
30. Farshbaf, Morteza. (2014). Investigating the role of museum in forming the culture of a society. National Conference on Architecture, Civil Engineering, and Development.
31. Foroughi, Neda., Fathipour, Zahra., Adibi Sadeh, Sanaz., & Darvishi, Hadi. (2014). Euler-Lagrange equation. The Second National Conference on Applied Research in Mathematics and Physics, Tehran, University of Applied Science.
32. Ghasemi Motlagh, Nasim. (2006). Problems with museums. Journal of Majles and Strategy.
33. Kargaran, Robabeh. (2016). World Museums and Iranian Arts. Thesis: Ministry of Science, Research and Technology, University of Shiraz.
34. Gombrich, Ernst. (2001). History of Tehran Art.
35. Maria Montes, Joseph. (2003). New Teachings. Translator: Akram Bahr-Oloomi
36. Mirzaei, Hossain. (2009). Comprehensive guide to Iran tourism in Ilam province. Publishing of Irangardan Tehran.
37. Nazar Zadeh (2015). History of museums in the world. Mahnameh Sakhteman Va Rooz Tajhizat (Monthly building and equipment of the day). No. 32.
38. Nafisi, Nooshi Dokht. (2001). Museumkeeping. Publications of the Organization for the Study and Compilation of Humanities Books.

39. Vaziri, Vahid., & Bakhshalizadeh, Mahdi. (2016). Architecture with Water. Garden water museum with a sustainable architectural approach. International Conference on Contemporary Islamic-Iranian Architectural Traditions, Ardabil, Mohaghegh Ardabili University.

English Sources

1. Ambrose, T. & Paine, C. (2018). Museum Basics: The International Handbook. Routledge.
2. Bedford, L. (2016). The Art of Museum Exhibitions: How story and imagination create aesthetic experiences. Routledge.
3. Bertinetti, M. Wawrzonek, D. P. Hill, P. K. Primrose, R. N. Thiele, H. Virr, A. ... & Price, A. M. (2015). U.S. Patent No. 9,072,848. Washington, DC: U.S. Patent and Trademark Office.
4. Brown, M., Levack, W., McPherson, K. M., Dean, S. G., Reed, K., Weatherall, M., & Taylor, W. J. (2014). Survival, momentum, and things that make me "me": patients' perceptions of goal setting after stroke. Disability and rehabilitation, 36(12), 1020-1026.
5. Brusiani, F. Falfari, S. & Bianchi, G. M. (2015). Definition of a cfd multiphase simulation strategy to allow a first evaluation of the cavitation erosion risk inside high-pressure injector. Energy Procedia, 81, 755-764.
6. Chao Yuan, Edward Ng, .٢٠١۴ Practical application of CFD on environmentally sensitive architectural design at high density cities: A case study in Hong Kong,Urban Climate, Volume 8,https://doi.org/10.1016/j.uclim.2013.12.001.
7. Colombo, E. Zwahlen, M. Frey, M. & Loux, J. (2017). Design of a glazed double-façade by means of coupled CFD and building performance simulation. Energy Procedia, 122, 355-360.
8. Engineer, A. (2015). Museum architecture: a new biography.
9. Falk John Howard. Dierking Lynn Diane (2013), Museum Experience Revisited. Walnut Creek, Calif.: Left Coast Press.

10. Gao, X. Wang, X. Yang, B. & Liu, Y. (2017, June). Design of a Computer-Aided-Design System for Museum Exhibition Based on Virtual Reality. In Chinese Conference on Image and Graphics Technologies (pp. 157-167). Springer, Singapore.

11. Ge, L., Dasi, L. P., Sotiropoulos, F., & Yoganathan, A. P. (2008). Characterization of hemodynamic forces induced by mechanical heart valves: Reynolds vs. viscous stresses. Annals of Biomedical Engineering, 36(2), 276-297.

12. Golding, V. (2016). Learning at the museum frontiers: Identity, race and power. Routledge.

13. Guo, W. Liu, X. & Yuan, X. (2015). Study on natural ventilation design optimization based on CFD simulation for green buildings. Procedia Engineering, 121, 573-581.

14. Hein, G. E. (2006). Museum education. A companion to museum studies, 340-352.

15. Henning, Michelle (2006), Museums, Media and Cultural Theory, Issues in Cultural and Media Studies, Maidenhead: Open University Press.

16. Homod, R. Z. (2018). Analysis and optimization of HVAC control systems based on energy and performance considerations for smart buildings. Renewable Energy, 126, 49-64.

17. Hooper-Geernhill Eilean (1999), (ed.) Museum, Media, Message. New York: Routledge.

18. Ito, K., Inthavong, K., Kurabuchi, T., Ueda, T., Endo, T., Omori, T., ... & Matsumoto, H. (2015). CFD benchmark tests for indoor environmental problems: Part 4 air-conditioning airflows, residential kitchen airflows and fire-induced flow. International Journal of Architectural Engineering Technology, 2(1), 76-102.

19. Khare, V. Nema, S. & Baredar, P. (2016). Optimization of hydrogen based hybrid renewable energy system using HOMER, BB-BC and GAMBIT. International Journal of Hydrogen Energy, 41(38), 16743-16751.

20. Kidd, J. Cairns, S. Drago, A. & Ryall, A. (2016). Challenging history in the museum: International perspectives. Routledge.

21. Koivisto, T. (2006). A note on covariant conservation of energy–momentum in modified gravities. Classical and Quantum Gravity, 23(12), 4289.
22. Lee, D. S. H. (2017). Impacts of surrounding building layers in CFD wind simulations. Energy Procedia, 122, 50-55.
23. LeVeque, R. J. (2002). Finite volume methods for hyperbolic problems (Vol. 31). Cambridge university press.
24. Loeblich Jr, A. R. & Tappan, H. (2015). Foraminiferal genera and their classification. Springer.
25. Macdonald, Sharon (2011), Collecting Practice, in A Companion to Museum Studies, Macdonald, Sharon (ed), Pp17- 33. Oxford: Wiley-Blackwell.
26. Modena, S., & Székelyhidi Jr, L. (2018). Non-renormalized solutions to the continuity equation. arXiv preprint arXiv:1806.09145.
27. Naboni, E. Lee, D. S. H. & Fabbri, K. (2017). Thermal Comfort-CFD maps for Architectural Interior Design. Procedia engineering, 180, 110-117.
28. Pierret, S., & Van den Braembussche, R. A. (1999). Turbomachinery blade design using a Navier–Stokes solver and artificial neural network. Journal of Turbomachinery, 121(2), 326-332.
29. Qunli Zhang, Yuqing Jiao, Mingkai Cao, Liwen Jin, 2017 . SIMULATION ANALYSIS on SUMMER CONDITIONS of ANCIENT ARCHITECTURE of TOWER BUILDINGS BASED on CFD, Energy Procedia, https://doi.org/10.1016/j.egypro.2017.12.690.
30. Radder, L. & Han, X. (2015). An examination of the museum experience based on Pine and Gilmore's experience economy realms. Journal of Applied Business Research, 31(2), 455.
31. Ricci, A. Freda, A. Repetto, M. P. Burlando, M. & Blocken, B. J. E. (2017, September). Urban comfort evaluation in an Italian historical district: the impact of architectural details in wind tunnel and CFD analysis. In URBANCEQ-2017: International Conference on Urban Comfort and Environmental Quality, 28-29 September 2017, Genova, Italy.

32. Sanborn, W. G. & Doyle, P. R. (2011). U.S. Patent No. 8,021,310. Washington, DC: U.S. Patent and Trademark Office.
33. Sera, F. Vicedo-Cabrera, A. M. Hashizume, M. Honda, Y. Schwartz, J. D. Zanobetti, A. & Gasparrini, A. (2018, August). Air Conditioning and Heat-Related Mortality in US and Japan: A Longitudinal Analysis. In ISEE Conference Abstracts (Vol. 2018, No. 1).
34. Sylvester, C. (2015). Art/museums: International relations where we least expect it. Routledge.
35. Taberner, A. L. (2016). The Rapidly Changing Legal Landscape Governing Museum Acquisitions of Cultural Property. In Cultural Property Acquisitions (pp. 21-38). Routledge.
36. Tong, Z. Chen, Y. & Malkawi, A. (2016). Defining the Influence Region in neighborhood-scale CFD simulations for natural ventilation design. Applied Energy, 182, 625-633.
37. van Hooff, T. Blocken, B. Aanen, L. & Bronsema, B. (2011). A venturi-shaped roof for wind-induced natural ventilation of buildings: wind tunnel and CFD evaluation of different design configurations. Building and Environment, 46(9), 1797-1807.
38. Wineman, J. D. & Peponis, J. (2010). Constructing spatial meaning: Spatial affordances in museum design. Environment and Behavior, 42(1), 86-109.
39. Yin, H., Guo, H., Lin, Z., & Zeng, B. (2017, July). CFD Simulation of Air-conditioning System in the Public Area of a Metro Station and Research on Energy-saving Operation Scheme. In 2017 2nd International Conference on Civil, Transportation and Environmental Engineering (ICCTE 2017). Atlantis Press.
40. Zeigler, T. Elias, E. & Barreau, C. (2017). U.S. Patent No. 9,694,651. Washington, DC: U.S. Patent and Trademark Office.

www.ingramcontent.com/pod-product-compliance
Lightning Source LLC
Chambersburg PA
CBHW070505200326
41519CB00013B/2720